Immigrants and the American Dream

Immigrants and the American Dream

★ *Remaking the Middle Class* ★

William A. V. Clark

THE GUILFORD PRESS
New York London

A Division of Guilford Publications, Inc.
72 Spring Street, New York, NY 10012
www.guilford.com

Printed in the United States of America

This book is printed on acid-free paper.

Last digit is print number: 9 8 7 6 5 4 3 2 1

Library of Congress Cataloging-in-Publication Data

Clark, W. A. V. (William A. V.)
Immigrants and the American dream : remaking the middle class / William A. V. Clark.
p. cm.
Includes bibliographical references (p.) and index.
ISBN 1-57230-880-X (cloth)
1. Immigrants—United States. 2. Middle class—United States.
I. Title.
JV6465.C58 2003
973.92′086′91—dc21

2003004906

Contents

List of Figures and Tables vii

Preface xiii

1 Immigration and the American Dream 1

Dreams and the Paths to Success 4
Defining the Dream 6
The American Dream and the New Immigrants 10
Viewing the Present through the Past 14
Trajectories of Success 17
Transformation of Places: A New Society 21
Reconsidering the Dream 25

2 Immigrants in the United States 29
Numbers, Flows, and Policies

Profile of the Foreign Born 31
Where the Foreign Born Live 37
Age Distributions and Family Growth 40
Socioeconomic Status and Human Capital 46
Policies, Immigrant Flows, and Global Connections 53
Observations and Implications 56

3 Making It in America 60
The Foreign-Born Middle Class

Defining the Middle Class 62
Becoming Middle Class 63
Evidence from Cohort Changes 70
Clues from Changing Generations 76
What Does It Take to Become Middle Class? 77

Outcomes in Regions and Communities 80
Timing and Economic Contexts 89
Summary and Observations 91

4 **Entering the Professions** 94
The Context of Upward Occupational Mobility 95
The Changing U.S. Labor Force 98
The Evidence for Professional Advance 104
Regional Variations 113
Professional Patterns by Ethnicity and Gender 115
Summary and Observations 122

5 **Reaching for Homeownership** 126
Homeownership in the United States 128
Trajectories to Homeownership 131
Outcomes: Homeownership Rates and Patterns 137
Housing Market Effects on Homeownership 145
Suburbanization and Spatial Assimilation 155
Who Owns?: Explanations of Success in the Homeowner Market 158
Conclusions and Observations 159

6 **Voicing Allegiance** 163
Naturalization and Assimilation 164
The Context of Assimilation and Citizenship 167
Naturalization and Becoming American 170
Naturalization Rates and Explanations 174
Political Representation, Immigrant Enclaves, and the Politics of Space 185
Observations and Conclusions 191

7 **Joining a Divided Society?** 194
Inequality and Its Outcomes 197
Progress and Inequality 200
Conclusions 206

8 **Reinventing the Middle Class** 208
Paths to the Future
People and Places 210
Constraints and Barriers 215
Population Growth and the Future 219
Concluding Observations 221

Appendix: Data and Data Sources 225
Bibliography 231
Index 247
About the Author 254

List of Figures and Tables

FIGURES

1.1. Economic status of the U.S. population: 1940–1996. 7

1.2. Characteristics of new single-family homes. 9

2.1. Number of foreign-born persons in the United States, 1990–2000. 30

2.2. Percentage of the U.S. population speaking a language other than English at home (population 5 years of age and over). 33

2.3. Foreign born in the United States by ethnic background, 2000. 34

2.4. Major origin of the foreign-born populations in the United States. 35

2.5. Proportion of foreign born by origin who entered the United States in the last two decades of the 20th century. 36

2.6. Distribution of the foreign born by states. 38

2.7. Percent change in the foreign born in the United States from 1990 to 2000. 39

2.8. Population of the United States by nativity, age, and sex, 2000. 41

2.9. Age–sex pyramids for immigrants from Mexico and Central America; China, Taiwan, and Hong Kong; the Philippines; and Russia and Eastern Europe. 42

2.10. Percentage of family households in the United Sates with five or more members by nativity and world region of birth, 2000. 43

2.11. Percentage of children whose parent(s) are foreign born by state. 43

2.12. Percentage of U.S. households with school-age children with at least one foreign-born parent. 44

2.13. Contrasting educational patterns for the native born and immigrants in the United States. 46

2.14. Population of the United States by nativity and educational attainment, 2000. 47

2.15. Population of the United States with at least a high school education by nativity and region of birth, 2000. 48

2.16. Immigrant median household earnings in the United States. 49

2.17. The historical context of U.S. immigration. 53

3.1. Growth in the middle class in the United States, 1980–2000. 65

3.2. Gains and losses in the middle class over time in the United States. 67

3.3. Middle-class households as a proportion of households: (a) total households and (b) for ages 18–34 and 35–64. 69

3.4. Changes in the proportion of middle-class households by cohort for the United States. 71

3.5. Changes in the proportion of middle-class households by cohort and ethnic origin for the United States. 73

3.6. Changes in the proportion of middle-class households by cohort by ethnic origin and region. 75

3.7. Percentage of U.S. middle-class households that are foreign born. 81

3.8. Growth in the middle class by state, 1980–2000. 84

3.9. Middle-class households as a proportion of all households by region. 88

3.10. The proportion of foreign-born households who are middle class by year and entry into the United States. 90

4.1. Change in the U.S. labor force, 1980–2000. 98

4.2. U.S. labor force growth of the foreign born by ethnic origin. 100

4.3. Change in foreign-born employment by industry in the United States, 1980, 1990, and 2000. 101

4.4. The foreign-born labor force in large-immigrant-impact states, 1980–2000. 102

4.5. Percentage increases in the native-born and foreign-born labor force for large-immigrant-impact states, 1980–2000. 103

4.6. Immigrant employment levels by occupational categories in the U.S. labor market, 1980, 1990, and 2000. 104

4.7. Size of the immigrant labor force by professional occupation. 105

4.8. Ethnic-origin composition of the immigrant professional workforce by origin in (a) 1980 and (b) 2000. 108
4.9. The immigrant professional workforce, 1980–2000. 113
4.10. Percentage of foreign-born workers in U.S. professional occupations. 114
4.11. Immigrant proportions of the professions for large-immigrant-impact states, 1980–2000. 116
4.12. Ethnic composition of the immigrant professional workforce. 117
4.13. Immigrant women as a percentage of the female professional workforce in the United States. 120
4.14. Immigrant women as a percentage of the female professional workforce by state. 123
5.1. Cohort trajectories in 1980–1990 for native born and foreign-born U.S. households. 132
5.2. Cohort trajectories in 1980–1990 for immigrant households. 133
5.3. Homeownership rates by period of entry into the United States. 135
5.4. Double cohort: age and time of arrival in the United States for Mexican-heritage males who are single-family homeowners, 1980–1990. 136
5.5. Homeownership rates for the foreign born in the United States and the four major-immigrant-impact states/regions. 139
5.6. Homeownership rates for households who arrived in the United States by 1980. 140
5.7. U.S. homeownership rates by age and ethnic background for foreign-born households. 141
5.8. Homeownership rates by age for foreign-born households for major-immigrant-impact states/regions. 142
5.9. Homeownership rates by age for Hispanic foreign-born households. 143
5.10. Foreign-born homeownership rates by low, middle, and upper income. 144
5.11. U.S. homeownership rates for the middle-income foreign born by age and ethnic origin. 144
5.12. Median housing values and affordability in two gateway cities in California. 149
5.13. Homeownership rates in two gateway cities in California. 150
5.14. Deviations in homeownership rates for selected ethnic origins in two gateway cities in California: percent deviations from U.S. foreign-born averages. 151
5.15. Homeownership in San Francisco and Los Angeles for the foreign born by age groups. 152
5.16. Homeownership by ethnic origin in Los Angeles by age groups. 153

5.17. Homeownership rates by period of entry into the United States for Los Angeles and San Francisco. 154
6.1. Naturalization levels in the United States by duration and ethnic background. 175
6.2. Homeownership rates for naturalized U.S. citizens. 179
6.3. Voting participation by period of entry to the United Sates and age (percentage of foreign born who voted in the November 2000 election). 181
6.4. Voting participation of naturalized U.S. citizens by education, occupation, and income. 181
6.5. Generational differences in voting participation, 1994–1998. 185
6.6. Growing Hispanic parity indexed by Spanish-surnamed elected officials in the Southwestern United States, 1960–1995. 188
6.7. Creating congressional districts and protecting minority (ethnic) voters. 190
7.1. Median U.S. household income, 1980–1999. 200
7.2. Household income variation by selected states and for age and household composition for the United Sates in 1997. 201
7.3. Inequality in U.S. and California incomes, defined as the ratio of the 75th percentile to the 25th percentile. 202
7.4. Income distributions by immigrant status in the United States. 206

TABLES

2.1. Foreign born (in Thousands) in 2000 by state with at least 500,000 foreign-born residents 32
2.2. Percentage of foreign born by state in 2000 (Ranked) 37
2.3. Educational attainment by immigrant status 48
3.1. Middle-class households in the United States by immigrant status and ethnicity 64
3.2. Ethnic middle-class households as a proportion of all middle-class households 66
3.3. Proportions of the U.S. population and proportions of the middle class by origin 67
3.4. Changes in the proportion of the 20- to 29-year-old middle-class cohort who arrived in the United States in 1980–1990 72
3.5. Changes in the relative proportion of second- and third-generation households who are middle class 77
3.6. Variables that are related to the probability of being middle class 78
3.7. Variables that are related to the probability of being middle class for selected countries of origin 79
3.8. The distribution of foreign-born middle-class households in 2000 82

3.9. Foreign-born middle-class households and their proportion of the total middle class by state 83
3.10. Proportion of middle-class growth that is foreign born 86
3.11. Percentage change in middle-class households by state 86
4.1. Employment in the U.S. labor force 99
4.2. Change in the native-born and immigrant labor force in 1980–2000 by industry in the United States 101
4.3. Composition of the immigrant workforce by origin in 2000 103
4.4. Foreign born in the U.S. professional labor force, 1980–2000 106
4.5. Professional penetration of occupations by the foreign born 107
4.6. Immigrants as a proportion of the professional workforce 115
4.7. Total number of professionals by state and selected countries of origin with more than 50,000 persons 118
4.8. Women in the professional workforce 119
4.9. Percentage of immigrants in professional occupations in 2000 who are women 120
5.1. Most common last names of recent homebuyers in the United States and California 127
5.2. Homeownership rates for foreign born by year for the United States and the high-immigrant-impact regions 138
5.3. Foreign-born homeownership in central cities and suburbs in the United States 156
5.4. Homeownership rates by residential moves in U.S. metropolitan areas 156
5.5. Predicting homeownership of the foreign born 159
6.1. Naturalization levels by age and year of arrival of immigrants in the United States 176
6.2. Naturalization rates by educational attainment of immigrants 177
6.3. Naturalization rates by race, ethnicity, and occupational status of immigrants 177
6.4. Naturalization rates by economic status, ethnic origin, and occupational status of immigrants 178
6.5. Political participation (percent voted) in the 2000 Presidential election by place of birth of the naturalized foreign born 182
6.6. Registration and voting by homeownership 183
7.1. Proportion of U.S. households with incomes in the lowest, middle, and highest quintiles 198
7.2. Measures of income inequality for households, 1980–2000 204

Preface

Two competing perspectives characterize the many studies of immigration to the United States: one celebrates the contributions of immigrants to their new societies; the other anguishes over the trying circumstances with which immigrants grapple every day. The positive accounts regard Latinos as the next Italians, certain to succeed through hard work and determination. The more negative would close the door to poor and unskilled immigrants from Mexico and Central America (Borjas, 1998). Those who would drastically curtail immigration emphasize the problems of integrating so many newcomers.[1] Those who advocate open doors see immigrants as the "resource" for the new century, focusing on the gains for the United States and the relief value for countries with too many workers and not enough jobs.[2] Clearly, strongly competing agendas inform these differing perspectives on immigration, among them the desire for cheap labor, or the need for skilled technicians, or even the humanitarian concern to afford opportunities for as many as possible.

In *The California Cauldron*, an earlier study of immigration to California, I sought a middle ground between extreme positions on immigration—between the celebrators and the worriers. At the same time it is not altogether obvious that there is a clear divide between these two groups, and if there is a divide, it is a complex one. Certainly one can celebrate the public gains from immigration and to immigrants individually, but one also can harbor concerns about the local outcomes and impacts in one community or another. Indeed, without a greater commitment on the part of the host society, some new immigrants will have a very hard time of it in their new country.

Economic hardship has often typified the lives of many newcomers to America. But not all of the new immigrants are poor and unskilled; in fact, the flow of immigrants and the foreign-born population in residence display wide variability. There is now substantial evidence of a growing middle-class immigrant population, both foreign-born (first-generation) and "home-grown" native-born descendants of earlier arriving immigrants. When we disentangle immigrant backgrounds and varying geographic locations, what can we say about that special group, those new members of the middle class who may be the precursors of other changes in the structure, composition, and socioeconomic status of the foreign born? Who are these new middle-class individuals, what are their occupations, and where do they achieve their dreams?

Achievement is a central and persistent facet of the American Dream—bettering one's status relevant to that of one's parents, or at the very least maintaining the family status. Entering a profession, becoming homeowners, engaging in civic affairs—these markers are deeply ingrained in the American psyche. These varied dimensions of immigrant success merit close attention, from occupational achievement to political participation. Immigrants, bringing little more than their strong ambition and a determination to succeed, often do indeed "make it" in America. They open businesses, enter the professions, and acquire homes. They gain elected office, and their voices echo beyond their local communities. In the chapters that follow I define the middle class more specifically. For now, I shall characterize success largely in socioeconomic terms, gauged by income, homeownership, success in the professions, and participation in the political processes.

The allure of continuing upward social mobility often leads to migration, and that migration brings "social sorting." Entrepreneurs of East and South Asian descent flock to Silicon Valley, Armenian businessmen to Glendale (in Southern California), and Russian merchants to New York City. Thus, the spatial patterns of immigrant success are also important in understanding how immigrants are faring. Upward social mobility is deeply rooted in the U.S. experience—and advanced through residential change. People change places to improve jobs, their lifestyles, and the houses they live in. Social mobility and residential change are inextricably interwoven. Because immigrants and natives move differentially, the outcomes for immigrants are different according to where they live, and in turn the outcomes for communities differ accordingly.

Not all immigrants succeed. Not all immigrants attain the American Dream, join the middle class, own a home, a car, and send their children to college. So who gains success, and where? Clearly, there is a very large "middle-class" Latino population in both California and Texas. Can we generalize more broadly to a geography of immigrant success? Are particular immigrants in certain regions more successful than their counterparts elswhere? If so, why? Does it relate to their innate human capital? Their education and training? When they arrive? An important aim of the book is to document where those success stories materialize.

In examining entry to the middle class, I focus on the fundamental issue of economic success: economic integration into the larger U.S. society. But the process of integration, and its timing, are interrelated with the general economic progress of American society as a whole. Since the 1980s there has been a growing concern about widening income inequality and the shrinking membership in the middle class. Thus, the questions about the trajectories of new immigrants are part of the larger debate about growing income bifurcation into the "have-mores" and the "have-lesses." Questions about the future of the middle class are questions about the health of meritocracy—the opportunity to rise with merit from humble beginnings. Now the question is, will the experiences of future generations, including the new immigrants, validate or call into question these underlying precepts?

In the following chapters, I examine the complex and sometimes conflicting processes of entry to the middle class. I emphasize both temporal and spatial changes, and I go beyond the narrow focus on income gains and losses. I seek to identify which new immigrants are advancing economically and where. (It is not just foreign-born engineers who are doing well, but a wide range of groups from different ethnic backgrounds.) We know that immigrants from some backgrounds are especially successful, and more so in some states and cities than in others.

Particular places differ widely in numbers of the foreign born who reside there and in the impacts that they have on those communities. The 2000 U.S. Census reports an increased spread of immigrants across the landscape of towns and cities in the United States. True, the majority of the foreign born still live in California, New York State, Texas, and Florida. But now large numbers of foreign-born households reside in nearly every state in the Union. Differences between old established immigrant centers like New York City and Los Angeles and new immigrant cities like Atlanta, Georgia, Orlando, Florida, and Las Vegas, Nevada, create different outcomes for the new immigrants. There are success stories in some locations, and other stories elsewhere. The patterns of middle-class growth are nevertheless remarkable in the large immigrant destinations.

Although I take a middle ground between the completely optimistic and the totally pessimistic view of immigrant outcomes, my perspective stresses the successful gains in occupations, industries, and housing markets. To the extent that public polices can nurture or reinforce these processes, they can be a part of an emergent diversified America. The assimilation of the new ethnic middle class in the 21st century will bring changing political agendas, and they will have a major impact on a wide range of policy issues. That is why we need to better understand who the new members of the middle class are, where they live, and what are the dimensions of their success.

Chapter 1 establishes the issues of social integration and the arguments about the relevance of the melting pot and economic assimilation. I am avowedly assimilationist, as I believe that it is through adaptation and incorporation that the new immigrants are likely to make the greatest progress to the middle class. Chapter 2 is an overview of the foreign-born population, which sets the current numbers in both historical and policy contexts. It is a framework on which, respectively, Chapters 3, 4, 5, and 6 on middle-class achievement, occupational successes, homeownership gains, and political participation are based. Given the sharpening indications of income polarization, even among native-born Americans, it is important to examine the new immigrant flows in the context of what is happening to the society as a whole. This is accomplished in Chapter 7. Finally, Chapter 8 summarizes the gains and constraints on the future paths of the immigrant middle class. I rely heavily on U.S. Census statistics, embellished occasionally by the immigrant stories that are the faces behind the figures. Although I am well aware of the increasing debates in geography and sociology about the value of qualitative research and appreciate how interviews and anecdotes can enrich any analysis, in the end I believe that the story of immigrant success is best portrayed primarily through a statistical analysis.

That so many immigrants do succeed is a testimony to their resilience and to the economy of the United States, which has absorbed nearly 10 million new immigrants in just the past decade. Despite problems for some and stunted progress in some locations, the American Dream of homeownership, upward educational mobility, and real economic gains for immigrants continues to flourish and, in doing so, validates the meritocracy to which immigrants have been drawn for centuries.

This book, then, is about the process of attaining the American Dream. Even though this notion may be intangible to many Americans, it is almost palpable to those who struggle to make the passage across the Rio Grande or on a cargo ship from China. These immigrants are drawn by the allure of upward mobility and the belief in its possibility. It is a translation of a "seeking-their-fortunes" psychology into migration and a new beginning. In the end, this motivation is stronger and more effective than the walls and gatekeepers and confused policy making about a U.S. immigration policy.

As always, I am grateful to numerous friends and colleagues who have patiently discussed the issues I address in this book and who have reviewed preliminary drafts. In particular, Peter Morrison and Eric Moore read the manuscript in draft and made excellent suggestions for reorganization and important new material. Some of the initial ideas were explored in a paper for the Center for Immigration Studies, and I am grateful for their support. I presented the core arguments at a recent Conference on Population Geographies, and the participants provided thoughtful commentary on my ideas. I would also like to thank UCLA and my department for sabbatical support during which time I was able to collate the background research for this project and produce the final manuscript. Chase Langford, the cartographer in the Department of Geography at UCLA, continues to translate my ideas for tables and charts into creative and striking presentations. Jeff Garfinkle, my longtime computer programmer, does more than just analyze the data; he asks probing questions that have helped refine the analysis itself. My wife, Irene, has listened as I worked out my arguments and brought her professional experience to bear on my prose.

NOTES

1. Peter Brimelow's (1995) *Alien Nation: Common Sense about America's Immigration Disaster* is one of the more forceful presentations of the negative consequences of immigration.
2. Julian Simon's 1989 book *The Economic Consequences of Immigration* argues strongly in favor of immigration as economically beneficial to the United States.

CHAPTER 1

★ ★ ★

Immigration and the American Dream

What is the American Dream to which immigrants are drawn? How does the dream attract so many immigrants—both legal and undocumented—to America, and how do they fare when they arrive? These are questions that are central to an understanding not just of the immigration flows of the last two decades but of the future influx from abroad, and of how American society will evolve and change in the coming century. Some eagerly celebrate immigration and the diversity it brings; others worry over the numbers and their impacts. One can be both a celebrator and a worrier, but the celebration story fits more neatly with the upbeat and positive views of America, and there is much to celebrate, even if it is tinged with some worries down the road.

This introductory chapter focuses on the dual nature of the dream as it is being realized in the 21st century. It addresses the varied facets of this dream—such as homeownership, education for one's children, and acquisition of material goods—and examines the varying paths new immigrants follow as they thrive and prosper.

A popular magazine article in the early 1990s asserted that unemployment is lower in Switzerland, owning a home is easier in Australia, attending college is likelier in Canada, yet dreams more often come true in America (Topolnicki, 1991). The headline was a teaser for a special issue of the magazine on the continuing importance of the American

Dream. The unabashed focus on material well-being ("despite what we've heard about our nation's decline—we still live better than anyone else") is an evocation of the successes of living in America. Is that dream still the lure for the dramatic flows of immigrants at the end of the 20th and the beginning of the 21st centuries? Is the dream of a new and better life that brought countless millions a century ago from Europe now bringing a new surge of migrants from Asia and Central and South America? Most important, are the new immigrants en route to achieving the American Dream, or are some becoming sidetracked into a backwater with few opportunities, where the struggle is simply one of staying afloat?

Dreams are intangible, and the American Dream is no less intangible than so many other dreams of our futures. Is the American Dream epitomized by Horatio Alger—clerk to corporate president, poverty to wealth, obscurity to professional distinction? Or is it exemplified more typically by such less glamorous outcomes as a steady job, a comfortable home, and a secure future for one's children? Does it still have relevance in our current society; does it still have the power to stimulate and excite, to generate tenacity and commitment? When one looks around the universities of the US,[1] as well as at the enterprises newly launched (whether high-tech "dot-coms" or low-tech gardening trucks plying their trade along the streets of Southern California), there does seem to be something at work. Both low-paying jobs and highly skilled occupations are filled by energetic people from other places around the globe.

Many commentators who explore the ideas of the American Dream speak of the collective dream embodied in the Bill of Rights (freedom from religious or political persecution), and to be sure this is still an important and enduring force in creating the context for immigration. Indeed, as the story goes, the immigrant who moved from Russia to New York was asked about why he had moved. Was it because of the housing? "No," he responded, "I couldn't complain." Was it because of the medical care? "No, I couldn't complain," he responded once more. Was it because of the job opportunities? "No, I couldn't complain," he said again. Then why, persisted the interviewer, did you immigrate? "*Here* I can complain," he replied. The story resonates with all who are motivated by the most basic desire—to the extent possible to have the freedom to be in control of one's own life.

However, most individuals are much more prosaic in their conception of the dream. The individual immigrant has always focused on material well-being and prospects for a better future, either in America or upon returning home with some tangible wealth. An early-20th-century

Italian immigrant celebrated his motivation to make money and return home:

> "If I am to be frank, then I shall say that I left Italy and came to America for the sole purpose of making money. I was not seeking political ideals. . . . I was quite satisfied with my native land. If I could have worked my way up . . . in Italy, I would have stayed in Italy. But repeated efforts showed me that I could not. America was the land of opportunity, and I came, intending to make money and then return to Italy." (Miele, 1920)

That dream persists in the late 20th century, for both the poor and the better-off. A recent newspaper report tells of a fashionable young hairdresser who, while doing well enough, wanted more:

> "I'm a fighter, I'm not satisfied with just getting by and that's what I felt I was doing here [in Mexico.]" . . . [I]t was the contrast of deterioration of life in Mexico with the constant reports of opportunities in the US which made up her mind. (LaFranchi, 1999, p. 1)

The dream was to do better. Contreras, the hairdresser was making what her friends called "a decent living." But there were "few prospects for improvement," and in the end it is that elusive search for improvement that is at the heart of the dream. Whether it is immediate gains for the individual or longer-term benefits for a family's young children, the prospects of moving up the ladder of success are all important. It is fashionable to decry the material gains of American society, even to "get off the ladder," but for many, and especially newcomers, the economic opportunities are paramount and are probably a greater part of the collective consciousness than we recognize.

A young computer engineer unknowingly paralleled the Italian immigrant from 80 years before. "Can I be frank?" asked Suman Kar, a 20-year-old senior at the Bombay Institute (a technology institute similar to Cal Tech or MIT), as he explained why he has accepted a job in Silicon Valley: "It's the money" (*New York Times,* February 29, 2000, p. A1). The job will pay nearly seven times as much as he would earn in India.

The dream is and was unabashedly material, nor was it much concerned with assimilation into a new society. It is the same dream that propels so many new immigrants today, the dream of improving their lot, of doing better. Repeatedly, media anecdotes of immigrant success recount the sacrifices the first generation makes to ensure second-generation successes. The native-born population may not resonate so fully with the American Dream, or even doubt its salience. Some of the native born are ready with an outright rejection of its mythology, but the immigrant population is embracing the opportunities offered by the American tradition of hard work, long hours, and often menial tasks. Who are these successful immigrants, where do they work, and where do they live? These questions will define the chapters that follow.

DREAMS AND THE PATHS TO SUCCESS

In a discourse on the American Dream, Hochschild (1995) suggested that it is a set of tenets about achieving success.[2] It is not just the outcome of a high income and a secure job; it is the enduring notion that even those who are poor and have limited skills can succeed. So many who are disadvantaged are still optimistic about their future. Here we have the two elements that are threaded through the American Dream, a belief that there is a fair chance of succeeding and ample opportunities to do so. Everyone has a chance, the opportunities are there, and hard work will be rewarded. Of course, it does not always work out so simply: skills and opportunities are not always perfectly matched; constraints and discrimination in the system prevent some from achieving their dreams; sometimes skills cannot be transferred from other societies. Even so, the enduring belief that effort will be rewarded is clearly a motivating force for so many of the new immigrants.

Attempts to define the American Dream have struggled with just how much the dream is spiritual and how much material. On the one hand, the dream emphasized a life which had the noble ends of freedom and self fulfillment—a life that was better, richer, and fuller. On the other hand, the American Dream included specific defining symbols: a house, a car, and abundant consumer goods (Galbraith, 1976; Reisman, 1980). In one of the more unabashedly material interpretations of the dream, a young couple sits gazing at the night sky and at vistas filled with a split-level ranch house, a sports car and family station wagon, and helpful home appliances (Calder, 1999, p. 3). Whatever its internal contradictions, the American Dream embodies both material well-being and the search for a life that is more internally satisfying according to every man or woman's ability. It is perhaps part of its enduring quality that it has this dual nature.

Those who have sought to interpret the American Dream have suggested that it has always been more than the search for material well-being. Even so, the evidence suggests that the search has been more material than not. Recent criticism of the notion of the American Dream has tended toward a rejection of the notion of upward mobility and certainly a serious castigation of the idea that the selfish and individual pursuit of the American Dream only generates overproduction and an orgy of consumption. To many, late-20th-century American society was one of heedless conspicuous consumption and little concern for its impact on the social and physical infrastructure. At the same time, commentators often fall back, albeit grudgingly, on the recognition that in some way the choice of democracy and the market economy is still a powerful force in creating our society. Even though it is clear that enlightened self-interest alone is not a panacea for the problems facing an urban society, there seems no other more persuasive ethic.

Whatever the confusion over the nature of the American Dream, it appears that the idea of relatively equal opportunities to pursue a wide variety of activities, including private economic interests, is an enduring force that is attractive well beyond national borders. The world is indeed critical of much heedless and thoughtless political behavior on the part of the United States as a nation. But as in Great Britain, Germany, France, The Netherlands, and all the democratic developed economies, other things being equal, the opportunities in such nations seem to outweigh the problems. The attraction of opportunities in a stable democratic society, even only a "relatively" caring democratic society, are powerful lures for many in poorer and less stable situations.

The American Dream embodies not only aspirations but also the avenues by which they can be realized. Without opportunities, dreams remain just that. But with opportunities the dreams can be realized, and it is the very fact that at least some dreams are being realized which is driving much of the immigration. In the minds of those pursuing it, the American Dream may be a loosely defined cluster of aspirations, but it clearly encompasses the chance to make money, to buy a house, and to ensure an education for the next generation. But it also has an element of individuality, of being able to do this on one's own in highly individual ways, unimpeded by authoritarian structures and to do it in a society, governed fairly, not corruptly. Of course, constraints are real and the opportunities may be tinged by inequality. However, it is some combination of personal freedoms and material opportunities that are at the heart of the enduring concept of American dreaming, and we may say French, German, Dutch, and British dreaming, because immigrants are seeking to enter those societies with the same intensity as they are seeking to enter the United States.

DEFINING THE DREAM

How can one define a dream? It obviously varies for different individuals and families. At the same time the discussion in the previous paragraphs suggested some common elements: a reasonable income, secure housing, and political freedom. The process of attaining the American Dream is in essence the process of becoming middle class, which encapsulates moving up the socioeconomic status ladder, becoming homeowners in (often suburban) communities, and participating in the political process.[3]

In this sense of the previous paragraph I am using middle-class status as a measure of success in realizing the American Dream. Still, there remains the problem of definition, as there is no official definition, no agreed-upon classification of those who are middle class and those who are not. However, even though there is no standard measure for the middle class, the concept exists in subtle forms, from casual conversation to television advertisements (Levy and Michel, 1986). In some ways it is easier to enumerate the concomitants of the middle-class lifestyle than to provide a precise definition. Clearly the concomitants include material goods, a home and at least one car, other consumer items like television sets, dishwashers, and personal computers, but also the funds to educate and raise healthy children and provide support for a comfortable retirement.

Income

Despite the lack of a generally accepted definition of the term "middle class," there is a very good working definition that we can use to guide our analysis. The University of Michigan Population Studies Center used a range of incomes linked to the threshold that defines a family in poverty. In their definition, the middle class ranges from 200 to 499% (or in other words from two to five times) the poverty line for a household of four.[4] The justification for this categorization is twofold. Using the poverty line as a control point ties the measure to a recognized basic support level for a family of four (a household). Equally important, the measure is relatively consistent over time and can be used comparatively across different censuses. Defining the lower level of the middle class at two times the poverty level excludes the poor and the near poor, but it also is sufficiently broad to capture both lower-middle-class and upper-middle-class incomes.[5] This measure fulfills the idea of tying the definition of "middle class" to a relative measure of income and generates a range of middle-class incomes. The definition for the late 1990s creates about a 40% middle-class distribution (Figure 1.1). The definition of 40% of all U.S. households as middle income is consistent with the broad findings of Levy (1998) and similar to the income findings of Leigh (1994).[6] Their income ranges were roughly in the region of $30,000 to $80,000 in 1997 dollars and are not especially different from the ranges we will use in the empirical analysis later in the book. The range suggested by Levy (1998) for middle incomes, $30,000 to $80,000, is quite similar to the 2000 range of $34,000 to $85,000 based on the above University of Michigan Population Studies Center definition.[7]

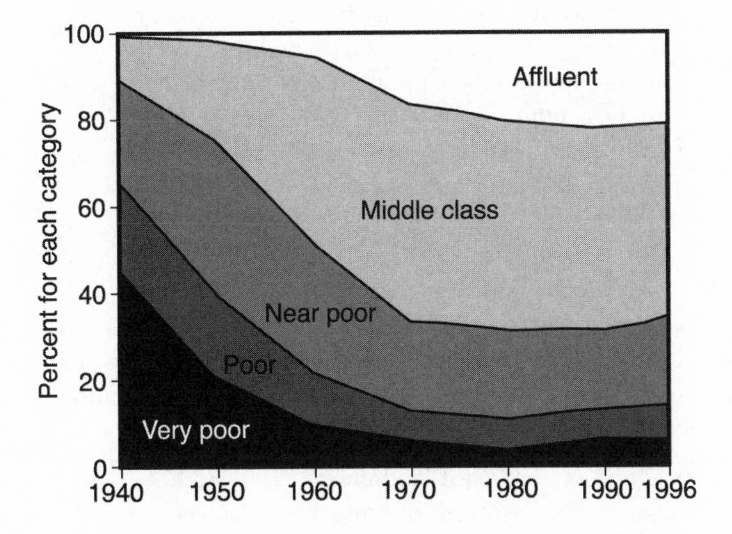

FIGURE 1.1. Economic status of the U.S. population: 1940–1996. *Source:* Analysis by the University of Michigan Population Studies Center of microdata samples from the U.S. Censuses of 1940–1990 and the March 1996 Current Population Survey, and modified from a figure in *Population Today,* vol. 25, November 1997.

Homeownership

The American Dream is more than just a dream of a good income. Another central dimension is homeownership. Owning a home is a core part of the dream, as it provides security and implies putting down roots and community commitment. Thus, income alone is an insufficient measure of the middle-class lifestyle, even though income is what affords access to the material goods which are the essence of the middle-class lifestyle (Levy, 1998). Buying a home is one of those critical purchases, perhaps the most critical purchase, and a central part of the American Dream. A fairly substantial literature notes that homeownership is linked to prestige and symbolizes "making it" in the United States (Ratner, 1996). But there are tangible reasons for making homeownership a central component of middle-class status as well—safety and autonomy, not to mention the financial and tax advantages of homeownership (Johnston, Katimin, and Milczarski, 1997). Thus, homeownership is an integral part of middle-class status.

Homeownership has taken on symbolic meaning beyond the value and assets of the home and is interconnected with the notions of upward and outward mobility—of increasing household assets and relocation to the suburbs. Moreover, the role of homeownership has become increasingly salient and central in the past half century. Before 1940, substantially less than half of U.S. households owned their own home, but since 1960 the average has climbed to about 66% in the country as a whole. Clearly, owning a house is now the norm and is a central part of the American and the middle-class lifestyle. Household surveys continue to reiterate the basic desire for homeownership and its pervasiveness across incomes (Heskin, 1983).

At the same time, a recent discussion ("The Muddle about the Middle Class," Population Reference Bureau, *Population Today*, Vol. 28, January 2000, p. 8) emphasized that varying living costs will determine who is middle class and that what makes up the middle-class lifestyle has changed over time. Nevertheless, the combination of an income range and ownership encompasses much that we think of as middle class, and that is the definition that will be central to my empirical analysis.[8]

Definitions and Perceptions

Most Americans identify themselves as middle class—either lower-middle class or upper-middle class—rather than working class or wealthy. The broad appeal of the middle class and its idealization has grown out of notions embedded in the American idea of equality and the ideals of upward mobility—the ideas that are central to the American Dream. There is a strong feeling that the United States is a nation, at least for white America, with near limitless opportunities for upward progress and continuing gains in material success. "Making it," defined in terms of a house, car, leisure time, and a secure retirement, is truly embedded in the American psyche.

At the same time definitions of the middle class are complicated because the basket of material goods that is regarded as symbolizing the middle-class lifestyle has changed over time. For example, a two-bath, two-car home is now closer to the norm for most middle-class households than the one-bath, one car home of the 1950s (Figure 1.2). It also takes more than one earner to create the middle-class lifestyle at the beginning of the 21st century. Households have changed in composition: two earners are common, and smaller households are the norm. This shifting economic and demographic context make it difficult to place boundaries on the middle class. But even though definitions are not straightforward and in the end are inextricably dependent on the exact quantitative measures used, the range I have suggested here is one that can be employed to examine the relative progress of both the native-born and the foreign-born population.

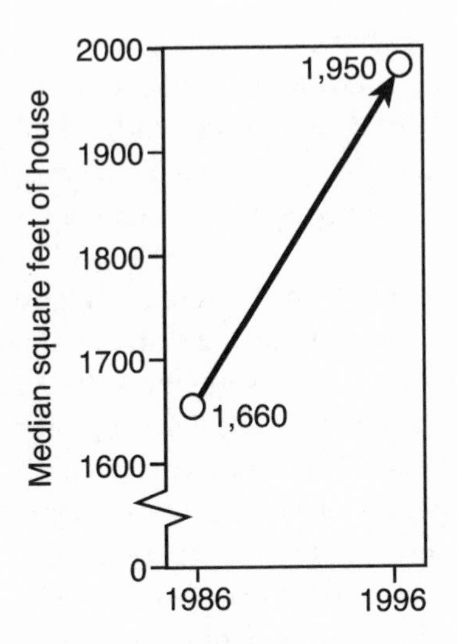

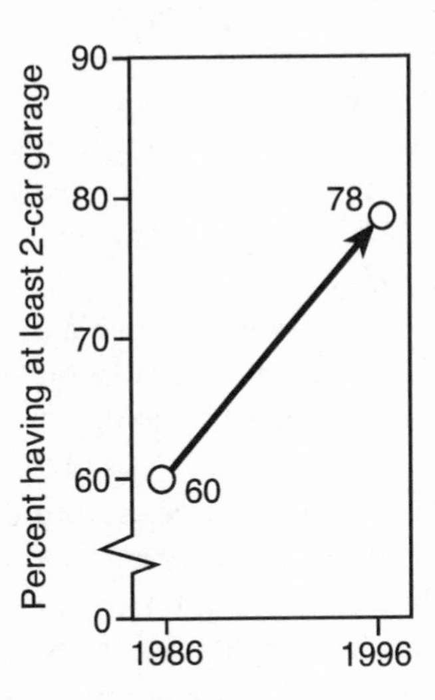

FIGURE 1.2. Characteristics of new single-family homes. *Source:* National Association of Homebuilders, *www.nahb.com.*

THE AMERICAN DREAM AND THE NEW IMMIGRANTS

The American Dream is an impressive ideology, luring people to America and thence to local opportunities in one or another place or region (Hochschild, 1995).[9] California in the 1880s was the dream of Charles Fletcher Lummis, city editor of the *Los Angeles Times,* and General Harrison Gray Otis, long the newspaper's publisher, and they sold the California Dream across the nation (McWilliams, 1973). The newspaper was the medium by which such men portrayed the opportunities and advantages of California, and the Union Pacific and the Atchison, Topeka, & Sante Fe railroads were the modes of transportation to bring people there. California as a destination embodied the dreams of internal migrants from the east coast. Now, however, the dream of California or the larger United States is conveyed not by newspapers, as it was in the late 19th century, but rather it is beamed electronically via satellites into the towns and villages of a visually interconnected world. American movies and TV programs provide powerful media for creating and vehicles for transmitting the images of American society worldwide—the allure of opportunities for individual advancement beyond a person's region of birth.

We know that a large part of the moving is done by a small part of the population (Morrison and Wheeler, 1978). Migration is likely to be self-selective, chosen by those "pioneer individuals" who are more willing to take risks, individuals who perhaps have a wider vision of the possibilities in unknown areas and who have a different perspective toward the future (Morrison and Wheeler, 1978, p. 80). These individuals are caught up in the image of "elsewhere" and perhaps its idealized possibilities, which have played a powerful role in the past and clearly continue to do so today.

The stories in the media and the dramas on the screen suggest that people everywhere should pursue their own hopes and dreams, and if the opportunities to pursue the dreams are not possible there, then move to where the dream can be pursued. Kerr (1996, p. 74) has suggested that the products of American culture are a "vast amorphous propaganda machine" which is capturing the imaginations of people everywhere. That may be an exaggeration, but it is certainly a powerful part of the imagery which is influencing the large-scale flows of boat people from Cuba, the substantial influx of immigrants smuggled from China, and the daily flows of undocumented migrants from Mexico. A common refrain bears out the hope of succeeding in America, and the lack of opportunities in their home countries often serves to reinforce the power of the American Dream.[10]

The hope of succeeding is relevant for all groups but especially for Latinos, whom *The Economist* (December 14, 1996, pp. 28–29) celebrated as the "new Italians"—coming to the United States without significant education and high-end skills, hard working, taking low-end, low-paying jobs, and integrating into the U.S. economy. A book entitled *The Americano Dream*[11] celebrates the successes of one Latino immigrant entrepreneur, Lionel Sosa, founder and head of the largest Hispanic advertising agency in the United States. The celebration of Latinos as the next Italians picks up a theme that will be a central element of later chapters in this book: the willingness to make severe sacrifices to achieve greater long-term goals. The parallel with earlier waves of Italian immigrants emphasizes the arrival of relatively poor and unskilled immigrants who eventually worked their way into the American middle class. Drawing the parallel suggests that the new Latino immigrants are more like the earlier waves of Italians than they are like the earlier waves of Jews or today's waves of Asian immigrants, who typically arrive with education and professional skills (what economists term "human capital"). The dream is the same even if the path to achieving the dream may be rather more complicated in today's changing global economy. (See the Appendix for a brief discussion of how the words "Hispanic," a Census term, and "Latino," a term often used in the media, are currently vying for acceptance by the public and in the research literature. Both terms are used in this book—"Hispanic" when census data are discussed.)

The Americano Dream unabashedly focuses on how to make it— how to use self motivation and how to transform the Latino cultural heritage into an asset, especially the strengths that come from family and hard work (Sosa, 1998). The concepts that we have seen as central to the American Dream are central to the *Americano Dream* as well. But there is a slight and important addition—hard work, individual reliance, *and* family guidance and ethnic identity, certainly Latino if not ethnic additions to the native-born perspective of making it with hard work and perseverance.

While *The Americano Dream* celebrates the success of its author, it is also a manual for immigrant success, a how-to book. The synopsis of the book notes that it will teach effective approaches to problem solving and, most important of all, an assertive, can-do attitude and ways to transform "your" cultural heritage into an asset that can become a viable tool for success. The marketing of the book emphasizes its value for anyone interested in starting a business or climbing the corporate ladder. Perhaps most interestingly, the book confronts the generational differences among Latinos, and especially the relationships between older and younger Latinos. Nor does the book shy away from the obstacles that may stand in the way of Latino success and, by extension, immigrant success. There are external and internal barriers to success. It is not only discrimination and societal barriers that may hold immigrants back, but also the internal barriers—lack of self-worth and feelings of equality—are equally critical. Thus, success involves more than simply acquiring human capital; it is acquiring a positive mind-set as well.

These examples highlight the continuing power and relevance of the American Dream. Success and its path may have different forms but still are part of the consciousness of the newcomers to the United States. They bring the same hope with which earlier waves arrived in the United States. And just as there were worries about whether the earlier waves were going to make it, those worries exist today. Contemporary observers were concerned about the concentration of Italians in the slum areas of large cities and in low-paying occupations (Nelli, 1983). Italians were contrasted with the thrift and self-reliance of the Germans and other immigrant groups from Northern Europe. Today's comparison of Asian and Mexican immigrants has a similar ring. While Asian entrepreneurs are often hailed as the integrated model minority (and it is true that they are more likely to be citizens), such comparisons, as we will see, are as flawed today as they were 100 years ago.

The American Dream and Assimilation

To an extent the American Dream, for the foreign born, implies melting-pot-type assimilation to American culture and values. But there is an increasing debate over assimilation, and assimilation has fallen out of favor and even into disfavor as an overarching terminology for the process of immigrant incorporation. Many social scientists have rejected the terminology as imposing ethnocentric and patronizing demands on minorities. Others have recast it to include multiple paths to incorporation in the new society—what is known as "segmented assimilation" (Portes and Zhou, 1993). Some have even suggested that assimilation is dead (Glazer, 1993). For nearly all there is general agreement that the paths to incorporation are hardly linear and that the process is more like a bumpy road than a smooth transition (Gans, 1997). The metaphor is now more mosaic than melting pot; indeed, it may be better to think of blending than assimilation.

There is no question that the paths to incorporation are quite divergent. Some groups are lagging, and at the other extreme many new arrivals often start out at parity with whites if not actually ahead of them (Waldinger and Bozorgmehr, 1996, p. 19). Monterey Park is but one of several well-established Asian middle-class communities around greater Los Angeles, and the suburbanization of immigrants is proceeding apace (Clark, 1998). Whether this stands assimilation theory on its head will continue to be debated, but it is worthwhile emphasizing that assimilation is more than just buying an expensive house in a middle-class suburb. It is a complex and multifaceted process.

There are attempts to provide a more nuanced discussion of assimilation and to rescue it from a premature grave. A thoughtful "rethinking" of assimilation theory while conceding the problems argues that assimilation as a concept is still useful (Alba and Nee, 1997). At its most general, assimilation can be seen as the "disappearance of an ethnic/racial distinction and the cultural and social differences that express it," as Alba and Nee (1997, p. 863) put it. They emphasize that it clearly cannot be viewed in the old normative terminology which favored an eradication of minority cultures. But they suggest that assimilation can still be used as a way of understanding the social dynamics of American society—that is, as a term for a process "that occurs spontaneously and often unintendedly in the course of interaction between majority and minority groups" (Alba and Nee, 1997, p. 827). For Alba and Nee assimilation remains a key concept for the study of intergroup relations.

Past discussions of assimilation invariably invoked the notions of the middle class as the norm or standard to which immigrants might aspire. For Gordon (1964) it was acculturation to the "middle-class cultural patterns of, largely, white Protestant, Anglo-Saxon origins" (p. 72), and the link to middle-class outcomes recurs in the discussions of assimilation which followed Gordon. Even Portes and Zhou (1993), in their discussion of segmented assimilation, identify acculturation to the white middle class as one of the possible paths of assimilation. Zhou (1997) in a reexamination of segmented assimilation also contrasts the paths of assimilation that can emerge when immigrant children are in contact with other poor minorities rather than the middle class. Because assimilation has always been linked to the notion of making it to the middle class, and because a major focus of the present book is on this progression to middle-class status, I believe that Alba and Nee's recasting and broadening of the concept of assimilation is useful for the discussions which follow. There is certainly an argument to be made that there were and are links between assimilation and seeking middle-class status. They occurred in the past, and (as later chapters will show) they are occurring today.

The debates about assimilation—its use and value—are likely to continue, but it is worthwhile making two points about its relationship to entry to the middle class. As DeWind and Kasinitz (1997) note, the discussion of immigrant incorporation is highly speculative and three or four decades is not a very long time in the immigration incorporation process. The interaction of the new immigrants in the coming decades with the changing U.S. economy and with changing social structures and political cultures, as well as consequent changes in the immigrants themselves, will likely produce outcomes that are not easily predicted. Previous waves of immigrants have made it and been incorporated into the changing American society: Many of the earlier waves of immigrants "are virtually indistinguishable on most economic and social criteria" (Hirschman, Kasinitz, and DeWind, 1999b, p. 130). It is quite possible that the same will happen for many if not all of the new immigrants and that the new groups may be equally indistinguishable in a new blended society.

Assimilation may be too easily and uncritically accepted, and just as easily and uncritically dismissed. We must draw power from the ideas without imposing a linear notion of assimilation. Even Gans (1997), certainly a critic of the assimilation concept, notes that in the long run the process of immigrant interaction in the new society may repeat many of the past findings of rapid acculturation and slower assimilation (Gans, 1997, p. 892). Along with Alba and Nee (1997), I regard assimilation as a useful concept for describing a process which is continuing, even if in more complex ways than in the past.

VIEWING THE PRESENT THROUGH THE PAST

In considering assimilation, it is useful to look back a century, as myth and distance have tended to cloud our understanding of those early migration flows. There is more in common between then and now than we may at first recognize. Immigrants in the early 20th century included educated Jews and Germans as well as poor rural farmers from Italy. Then as now, immigrants were drawn by the prospect of jobs that could provide money to send home to their families. Contemporary observers and later analysts documented the seasonal nature of the migratory flows from Italy and the heavy remittance transfers back to Italy (Nelli, 1983).

The early immigration from Italy was made up of largely unskilled working-age males, echoed a half century later in Mexican immigration to Southern California, before the immigration laws and the Immigration Reform and Control Act of 1986 (IRCA) changed the dynamic.

As it is today, the American Dream has long been a motivating factor in earlier waves of immigration to America. Fascinating remnants of that dream from earlier eras can be discerned in particular settings. Rosedale, Mississippi, a community in the heart of the Mississippi Delta, is one such setting. Here, Chinese immigrants found a niche in grocery stores and service activities. Originally plantation workers, certain Chinese found other outlets for their talents. Wong's Food Market willingly served blacks in an era of segregation when others refused. A half-century later, their market—and elderly Wongs—remain, entrepreneurs whose children have moved away to higher rungs on other ladders—in San Francisco, New York, and Los Angeles.[12]

Contemporary immigration to the United States is indeed at levels which are similar to those in the first decades of the 20th century. It generates the same processes of social integration and upward mobility. And, as in the earlier flows, the current waves bring a diverse mixture of poor and better-off immigrants. The United States as a whole has absorbed more than 20 million legal and undocumented immigrants in the past three decades (Clark, 1998; Smith and Edmonston, 1997), with little to suggest that the levels of influx will decrease anytime soon. As is well known, the changes in the size and composition of the flows were initiated with new immigration legislation. The Hart–Cellar Act of 1965 changed the terms of entry and, by emphasizing the mixture of skills and family reunification rather than country quotas, reshuffled the origins in favor of Asia and Mexico and Central America rather than Europe. Although the change in immigrant policy was designed to shift the emphasis to a skill-based quota system, other changes opened the immigration door for immigrants who were less skilled than previous waves of immigrants.

Migration to the United States in the early part of the 20th century was heavily labor-market driven but not solely so. Some immigrants came as religious and political refugees. However, jobs were important, and when there was an economic downturn in the United States, the laborers returned to their home countries in Europe (Nelli, 1983). Migration was sensitive to employment conditions. The demand–pull migration flow of the early 20th century was replicated during and after World War II in the Southwestern United States with the shortage of agricultural workers from 1942 to 1964. (The same phenomenon occurred in Germany in the postwar years.) But unlike the earlier period when immigration slowed as the economy slowed, the new immigration had a self-perpetuating dynamic, fueled by supply–push factors and the persistence of "beaten path" networks established by the earlier flows. The expanding population of Mexico and the lack of jobs generated a continuing flow of job seekers, who crossed the border into the United States any way they could. The flows were often highly focused spatially, both in their origins and their destinations. Added to the job flows were the refugee populations from Southeast Asia and destabilized Eastern European nations.

The enduring networks established during the era of permissive labor migration practices of the 1960s and 1970s set up the information networks linking origins and destinations. These networks contained information not only about the job opportunities but about ways of getting to the United States and where to find a safe haven. All this was the basis for flows of family members in the 1980s and 1990s. Several studies of Mexican communities have documented the initiation and perpetuation of migration between Mexico and the United States (Massey and Espinoza, 1997; Massey, Goldring, and Durand, 1994). Despite recessions, the flow of legal and illegal migrants has continued,[13] a trend that is relatively new in immigration globally but which has rapidly assumed a significant proportion of all population movements worldwide. Earlier waves came by boat and were mostly processed for permanent or temporary entry. Now immigrants come by foot and by air as well as clandestinely by sea. The illegal flows have become an issue in a time of refocused concerns on the nation-state and the role of law (Hollifield, 1996). In fact, the latest data suggests that the number of illegal immigrants in the United States may be more than 8 million, nearly 3% of the total population (Passel, 2001; Warren, 2000).

The expansion of civil rights legislation to encompass minority groups other than African Americans provided a more receptive climate for foreign-born groups than had been present during the earlier waves of immigrants. Judicial activism, the rise of immigrant advocacy groups, and the advent of numerically large ethnic minorities in communities in the United States has further contributed to expanded rights for foreign-born ethnic groups (Cornelius, Martin, and Hollifield, 1994). The confluence of the demand for labor and the emphasis on the rights of immigrants to have the same protections and privileges as those of the native born have certainly made immigration a less traumatic experience than it was for those who arrived in the past without protections. Thus, immigration has become much more than simply a labor-supply issue; as the numbers have increased, the issue of how the immigrants will fare and assimilate has become again a central part of the public discussions of immigration. Will they in fact assimilate in ways which are similar, or at least appear to be similar, to the patterns of assimilation of the earlier waves of immigrants from Northern and Southern Europe?

TRAJECTORIES OF SUCCESS

Invariably, immigrant success evolves over considerable time. In the past immigrants usually arrived poor, and the classic path was a slow trajectory extending across several generations to a more secure and successful position in the new society. Each cohort does better than its predecessor cohort. Even then, success was not guaranteed and required a passage through perilous and unsafe labor conditions. It was the sweatshops of late-19th-century and early-20th-century lower Manhattan, difficult as they were, that provided entry-level jobs for those who had few if any skills. The classic path is one in which the immigrant arrives poor and with few skills, precariously gains a foothold on the first rung on the ladder and slowly moves up. But, as demonstrated later in this book, it is only one of the possible paths at present. At this point, immigrants are arriving who may be considered "already" middle class, some of whom have significant levels of education and important previous professional training—the human capital mentioned earlier.

The social mobility of the past waves of European immigrants has been extensively documented in the sociological literature. There has been significant convergence in economic status, educational levels, life chances, and residential patterns between the descendants of the earlier waves of European immigrants and the original American settlers (B. Duncan and O. D. Duncan, 1968; Hirschman, 1983; Lieberson and Waters, 1988; Neidert and Farley, 1985). As documented in the seminal contributions of Lieberson and his colleagues (e.g., Lieberson, 1985; Lieberson and Waters, 1988) for white immigrants from Europe, the differences between them, their descendants, and the original American stock have largely vanished in the several decades since the waves of migration in the first two decades of the 20th century.

Much of the current debate about the future paths of the foreign born in the United States revolves around whether or not the classic path of social mobility is still accessible to new immigrants. Within the debate there is a subdebate about whether the new immigration is contributing to the polarization of society into rich and poor, and the shrinkage of the middle class. Part of this concern is whether or not there is a bifurcation within immigrant flows as well, into rich and poor newcomers to America.

All of these questions have generated considerable confusion and heated debate. Do the new immigrants have a chance of making it into the middle class, and who is making it? Will the individuals who arrive at the bottom remain trapped there? Some evidence suggests that the old path is still open. Gottschalk (1997) shows, for the population as a whole, that of those in the lowest-income quintile (the bottom 20%), more had progressed out of the lowest quintile than were still in it 17 years later: 58% had advanced above the lowest quintile, 23% had moved up one quintile, 14% had made it up two quintiles, 13% up three, and 8% had reached the top quintile. Research demonstrating such fluidity is at the heart of arguments about immigrant success. Clearly, a remarkable proportion of immigrants does move up in the classic pattern, even though many stay behind. The earlier processes were the same: not everyone made it, and certainly not in one generation.

Specific studies of Latino immigrants in Southern California paint a similar picture. Myers (1999) examines the changes in particular age groups arriving at about the same time, finding substantial evidence of upward mobility. Over time, immigrants move out of poverty, from the city to the suburbs, and become homeowners. The data showed that Latino and Asian immigrants often escape poverty over time and gain access to suburban homeownership—exactly the process we would expect of new immigrants. Rodriguez (1996) tells a mirroring story of entry into the middle class. Latino immigrants are doing better over time, more are in the middle-income ranges, and many have joined professional occupations. For many Asian and Middle Eastern immigrants the findings are even clearer.

Trajectories of success are often measured in terms of social integration, of assimilation to the host country mores. Are the new immigrants assimilating to the host country patterns? Are they able to integrate into the economy and the society? These questions are at the heart of much recent research claiming evidence of Balkanization and separation (Frey, 1996), and of segmented assimilation, which was discussed earlier (Zhou, 1997). To reiterate an earlier observation that remains true of the current scene, it is important to recognize that it has always taken time for immigrants to move into the mainstream (Rodriguez, 1996). Each new wave of immigrants subtly changes what it means to be an American. For Rodriguez, the question is not whether immigrant groups have cut their ties to their homelands but rather whether they are putting down roots in the United States.

By such measures of "rootedness" as citizenship, homeownership, language acquisition, and intermarriage, the evidence favors assimilation and trajectories that are following the patterns of previous immigrants. In the past, immigrants were slow to become citizens, but that seems to be changing, even for Mexicans who were traditionally much less likely to naturalize.[14] Homeownership rates are rising and are extremely high for some immigrants groups (Clark, 1998). Moreover, the longer residents are in the country, the more likely they are to be homeowners. Second- and third-generation immigrants are very likely to speak English at home and their intermarriage rates are high (Allen and Turner, 1996; Clark, 1998; Rodriguez, 1999).

Sometimes success comes for the first generation. Mee Moua, a lawyer and lobbyist, left the Laotian highlands with her family when she was a child, part of the Hmong (Montagnard) refugee migration of the early 1970s (*New York Times*, February 2, 2002, p. A13). Now she is the first Hmong elected to a state legislature. Her election to the Minnesota Senate is another signal of the way in which immigrants transform themselves to citizens and participants in American democracy and American society. Ms. Moua is clearly a member of the middle class, professional, homeowning, and now not only a political participant but a policy maker as well.

There is also strong evidence for direct additions to the middle class. Migrations in the late part of the 20th century have in many cases been of people who have more skills and greater education than the population of the country they are from. Among immigrants from India are many who are skilled engineers and managers. Similarly immigrants from Malaysia, Thailand, Hong Kong, and China are often skilled and highly qualified. Middle Eastern and Korean immigrants who have become the entrepreneurs of small and not-so-small businesses came with both human capital and financial resources. There are middle-class professional flows from Mexico and Central America, countries which are often identified as the origin of low-skilled and poor immigrants.

Bifurcated Flows

At the same time, the bifurcation of the immigrant flows cannot be ignored. It is a function of the changing economy, growth in high-tech industries and low-skilled service jobs at the same time. In the middle 1990s, the number of low-paying jobs (under $15,000 annually) has grown by about 4% a year—twice as fast as all other jobs. At the same time, a quite significant expansion of jobs has occurred in the technology sector, a growth which is still occurring despite the recent downturn in the economy.

The bifurcation of immigration into flows of "haves" and "have-nots" partly reflects the increasing number of refugees who are arriving in the United States. Many of these newcomers were already poor in their native countries and were relatively low skilled. In addition, the refugees are coming with few resources. The evidence confirms that many refugees are poor; in California at least, "long term welfare dependence is the norm for many refugees" (Barnett, 1999, p. 3). Approximately a million refugees have been admitted in the last 10 years. In addition, many refugees have arrived temporarily as a result of natural disasters or civil wars and other destabilizing political events. These temporary refugees are likely to be granted permanent status. The temporary residence for the large influx of Central American refugees after Hurricane Mitch in 1998 is due to expire but will probably be translated into permanent status. Similarly Kosovar Albanians and Liberians are likely to have their refugee status changed. All of this reiterates the diverse paths by which immigrants attempt entry to the United States and to grasp an opportunity for upward mobility. It is worth recalling that these foreign-born groups are entering in a wholly new context and therefore we should not simply group the refugees with other economically motivated immigrants.

Refugees and poor, low-skilled workers are the ones most likely to have real difficulty making social and economic gains. However, the data show that the poverty rate for all Latinos in Southern California increased only slightly and that poverty declined among those who arrived in the decade of the 1970s. Clearly, an influx of newcomers with high poverty levels is what has pulled the average down (Myers, 1999). Second- and third-generation Mexican American children were less likely to be in poverty. Specific groups such as Vietnamese children were also less likely in general to be in poverty (Oropesa and Landale, 1997).

About a third of all recent Latino immigrants live at or below the official poverty line. The recent immigrants, legal and illegal, have nowhere to start but at the bottom of the economic ladder. In addition, the flows generated by family reunification are continuing to add to the poor population. There are now more poor Latinos in the United States than there are poor African Americans. But again the story is not without its positive spin. The new Latino immigrants by and large are in the workforce: Many of them have found low-skilled, low-paying jobs, a condition that is not unusual in the immigrant experience. The issue, as always, concerns the avenues of upward mobility. The very size of the poor population, and the continuing supply of additional low-skilled poor immigrants, may create an underclass for which there is no way upward (Clark, 1998).

The flows of immigrants with significant accumulations of human capital are a direct response to the restructuring of the American economy to emphasize both high-level and basic services. In the former, bright creative minds from anywhere in the world can find jobs in the financial markets and computer software/engineering firms that have sprung up to operate the late-20th-century economy. These same workers, often immigrants themselves, need a service population to tend lawns, care for children, park cars, and wash dishes in the new restaurants that have sprung up to cater to this new high-end population. In addition, changes in just-in-time manufacturing has re-created the sweatshops of offshore companies in urban California garment districts. Together, these bifurcated flows are transforming the immigrant process and the places they settle.

TRANSFORMATION OF PLACES: A NEW SOCIETY

Immigration transforms those who embark upon it; it also transforms the places where they settle. The latter changes, in the end, are no less significant or noteworthy than the former. Together, the transformations of people and of the locales they inhabit are what is altering American society. Many immigrants will make it and move up and enrich the neighborhoods into which they move, though some will find difficulty in moving up to better-paying jobs and are likely to remain clustered in inner-city barrios and ghettos.

Earlier anecdotal reports, buttressed by U.S. Census 2000 data, paint a picture of notable—sometimes dramatic—local change through immigration. Immigrant flows are now branching out beyond entry-port states like California, New York, Florida, and Texas to numerous locales that offer these new Americans opportunities to thrive and prosper. They include small and medium-size cities in the Midwest and the South. Locales as different as Las Vegas, Nevada, Lexington, Kentucky, Nashville, Tennessee, and Fairfax County in Virginia are experiencing significant transformations. Las Vegas has had a 139% growth in the Hispanic population, and that growth is replicated in a large number of small and medium-sized cities across the country. Immigrants are not just arriving and staying in the gateway cities; they are moving from Los Angeles, Dallas, Phoenix, and San Francisco to cities where there are perceived opportunities. In Las Vegas, Lilia Guzman learned that even a kitchen helper can lead a middle-class lifestyle, including health insurance, vacations, and homeownership (*Los Angeles Times*, November 30, 1999, p. B-1). The Korean Business Directory in the Washington, DC area lists 560 Korean-owned businesses in Annandale, Fairfax County (*Washington Post*, May 16, 1999, p. A1). To be sure, the success stories are always attractive and not all moves end happily, but these two short anecdotes exemplify the changes that immigration is bringing to the nation as a whole and not just to the high-immigrant-impact states.

Small communities like Dodge City, Kansas, Amelia, Louisiana, and Georgetown, Delaware, have seen a dramatic increase in the number of immigrants. Onetime seasonal workers looking for more permanent work were willing to take low-paying jobs in chicken-packing plants and have moved from being migrant seasonal workers to becoming permanent residents. The growing immigrant population has generated opportunities for small businesses to serve the immigrant community. Drawn by increasing numbers of fellow immigrants, they set up groceries to provide familiar foods or services in a familiar language. It is these entrepreneurial activities which begin the long process to integration, acculturation, and the middle class.

Often the newcomers are repopulating communities that were in decline. Nearly 10% of the population in Utica, New York, are former refugees, perhaps the highest proportional concentration in the United States. An eclectic mix of Bosnians, Russians, and Vietnamese have come to Utica, drawn to jobs not in the old factory-based economy of General Electric but to those in countless small businesses engaged in telemarketing, check processing, and telecommunications. Many of the new immigrants originated in Eastern Europe, speak passable English, and often are well educated. The influx of refugees and other immigrants has strengthened local economies, increased the tax base and often revitalized inner-city run-down housing (*Wall Street Journal*, March 8, 1999, p. 1).

Examining *only* immigrants or *only* places yields an incomplete picture of the immigrant process. It is the intersection of opportunities localized in places and of immigrants willing to take risks that, in combination, generates so many variations on the common underlying transformation and immigrant assimilation process. Immigration and assimilation are dynamic processes. Given the relatively high levels of mobility in American cities and communities, those who are more successful venture beyond their immigrant neighborhoods. If the arriving immigrants who replace the departing ones happened to be poorer, that community registers an increasing poverty rate, even though the individual migrants may be doing much better over time. However, the "place" effect is real: poorer people in residence may lower tax bases and so diminish available resources to help the new immigrant population. The issue to consider concerns the difference between outcomes for places and outcomes for people. If poor but employed immigrants move into a neighborhood and crowd together in marginal housing, the local community may register a worsening of local conditions. But these new individuals, perhaps unemployed or marginally employed before, experience real gains in their lives. The place may be worse off, but the in-migrants are better off. To the extent that the immigrants can in turn improve the neighborhood, there are gains for the community as well as the newcomers. In the latter case, the influx of new immigrants can mean revitalization and a growing tax base.

In other instances new immigrants settle immediately in the suburban communities of the large gateway cities, exemplified in those of Southern California as well as of San Francisco, Dallas, and numerous other large metropolitan areas. As older native-born white owners retire and move to retirement villages and warmer climates, immigrant families are purchasing their houses. The more affluent groups have created suburbs within the suburbs (Li, 1998). The data on suburbanization in California reveals that the foreign born who have more education and are citizens and professionals are very likely to have moved to the suburbs of San Francisco and Los Angeles (Clark, 1998). The 2000 Census data report significant growth of foreign-born ethnic groups in nearly all medium-sized cities in the United States. Nor is this a phenomenon occurring only in the United States. A detailed study of Toronto, Canada's largest city, reports on South Asian, Jamaican, and Filipino families moving to the suburbs north and west of central Toronto (Bourne, 1996).

Support Services and Local Contexts

Opportunities for upward mobility exist in a wide variety of contexts both local and national. However, the concentration of new immigrants in a few neighborhoods in a small set of gateway communities, the entry ports for the upward mobility they seek, requires local support to facilitate upward mobility. Without local government services and the support of religious and other nonprofit organizations, new immigrants often have difficulty making the upward shifts they are striving for. In essence, the investment in education, job training, and health care is providing the basic resources to increase the human capital of the new immigrants. It is through the support of local governments that immigrants can advance toward mainstream incorporation.

Places serve not only as entry points but also as homes and communities, which are themselves culturally transformed. As the immigrants revitalize existing businesses and add new ones, they turn their dreams into reality. At the same time, there is often an undercurrent of resentment when, for example, signs may only be in Korean, Persian, or Thai. Clearly, it is in the best interests of the new businesses to make their sites more accessible by using bilingual signs. But new immigrants have traditionally served their own ethnic groups first, as a means to the successful creation of a path to financial security—in essence to the middle class.

With changing neighborhoods come changing social and cultural traditions—Mexican soccer and social organizations, and Asian Mah-Jongg clubs—which in turn make our communities more like home for prospective migrants. Perhaps the most welcomed is the proliferation of new ethnic cuisines. And, as people's tastes broaden and cultures intersect over food, the cultural landscape itself gradually transforms. A few blocks from Los Angeles City Hall, Jean Han, the Korean owner of a tiny fast-food restaurant, *The Kosher Burrito*, serves up an eclectic mix of food to an equally mixed clientele. And, as the cultural landscape changes, so too will the political landscape, as candidates running for elected office have to consider a diverse population with different needs from the formerly majority white population. The ferment between immigration and social and cultural change and what was once unique to some large inner-city communities will soon be commonplace in communities across California and the nation. The look and the feel and the issues that have been central in multiethnic counties like Los Angeles will become the look, feel, and issues of Fresno, Stockton, Modesto, and Visalia in California, and soon Rockford, Illinois, Peoria, Indiana, Syracuse, New York, and Wilmington, Delaware.

Perhaps more than any other cultural phenomenon, the emergence of soccer is a metaphor for the impact of Latino immigrants on local communities in the United States and especially in California. In the latter half of the 1990s, soccer games held in the Coliseum in Los Angeles, built originally for the 1932 Olympics, host to the 1984 Olympics, and home of the USC football team, have transformed the sport in Southern California

(O'Connor, 1999). It is immigrant driven. Mary Price (2002) describes how soccer has created Latino cultural spaces in Washington, DC. In Los Angeles soccer drew more fans in 1998 than the Raiders final National Football League season there in 1994. But even more important than the professional soccer teams in the big stadiums are the countless "pickup" Saturday morning and afternoon games in countless small urban parks in Los Angeles, San Jose, Dallas, and Fort Worth. The millions of young teenagers playing in the soccer leagues have already changed the cities and towns that were once devoted to American football only.

Where only two decades ago the transformation of society occurred in particular locations and only slowly diffused across the country as a whole, that process is accelerating across the nation. Obviously the greatest changes occur where most immigrants settle, so the changes appeared first in the large cities, in New York and Los Angeles. Now, those changes are diffusing across the nation, affecting communities large and small, from the Midwest to Appalachia, as new migrants branch out across the nation in search of the American Dream.

RECONSIDERING THE DREAM

The following chapters focus on how new arrivals in the nation as a whole and in large immigrant population states in particular acquire education and human capital, and eventually become American citizens and enter the middle class. But my intent is to go beyond mere statistics to the less easily quantified allure of the motivating force, the immigrant dream of both material gains and personal freedom. Not all the immigrants, of course, surmount the obstacles of poor schools or manage to find well-paying jobs; indeed, many immigrants do appear to be having great difficulty in make the transition to the middle class. Are they less likely or more likely than the native born to make that transition? That question is central part of the focus of the chapters that follow.

There are two kinds of immigrant stories in the popular media. One is about new immigrants who work hard and whose children are the high school and college valedictorians. Then there are the tales of poor families who suffer hard luck and misfortune. Not surprisingly, the media is attracted to the success stories, to the stories that fit with the image of the American Dream—hard work, perseverance, and success. Stories of immigrant single parents laboring in sweatshops are less engaging because success has not yet materialized out of hard work and perseverance. In the long run, they or their children may yet be successful. An important goal of this book is to attempt some assessment of generational success.

A recent positive media story headlined "Zero Down, Hard Work and Dreams That Came True" (*Los Angeles Times,* July 17, 1999, p. A-1) captured the most positive story of immigrant striving. An El Salvadoran family, the Garcias, the father, the mother, and five children, arrived in California a decade ago, fleeing the war in their home country. Now, Manuel Garcia operates an independent truck, they own their home (and other residential property), and all five children are in college or college bound. Clearly this story epitomizes the American/California Dream: a tale of an intact, hardworking, upwardly mobile family who had the abilities and skills to use the opportunities that were available to them. But, Manuel has not taken a vacation in 11 years, and the birthday and graduation celebrations are far from ostentatious. For this family, the difference was between their home country where it was living day by day or the chance to make something in safety and security. Those intangibles are as much a part of the dream as are cheap credit and rewards for risk taking.

The Garcia family had one thing in common with many of the new immigrants. They had family in the United States and could use that initial contact as a start in the long process of becoming American. But for another immigrant, Miguel DeLeon, the lack of family contacts did not deter upward mobility. However, Miguel had another advantage: some education, which enabled him to enter a managerial position. When the business was going to fail, he fell back on that other immigrant characteristic, the willingness to take risks. He took over the business with the help of small business loans and sacrifices by his fellow employees. The process is never smooth, but for those with some human capital, a willingness to work hard, and the know-how to access loans or credit, the American Dream is possible. Immigrant success might be defined as having a good job, an adequate income, buying a house, and participating in society—in sum, becoming middle class. These "successes" define the chapters of this book and constitute its organizing theme. As Mr. DeLeon laughingly commented in a *Life and Times* television interview (Oct. 22, 1998) in Los Angeles, the American Dream means a house, a car, education for the kids, and a dog and a cat. While this book is not about the dog and the cat, it is about the house, education, and making it to the middle class.

The American Dream remains a pervasive idea if only because people want to believe it. We want to believe that anything is possible, that wealth is a function of brains and hard work rather than influence or inheritance, and that American society as a whole provides the milieu in which this can happen. The inspirational tales of immigrant success have found a place in the hearts of those that are already here, and they are a potent force in generating the continuing flow of new arrivals. They are all focused on the chance of joining the American middle class.

NOTES

1. Nearly half my colleagues in the Geography Department at UCLA are foreign born, from Canada, China, Germany, Hong Kong, Trinidad, and the United Kingdom. In several of the science departments at UCLA the proportion of foreign-born professors is even higher.
2. Although Hochschild's book (1995) is an excellent discussion of the power of the American Dream, the book is in fact quite critical of the concept itself. The book emphasizes the flaws in the dream, especially for African Americans.
3. It is important to drawn a distinction between the use of the term "class" in the American context with that of British use. American usage tends to emphasize socioeconomic status as a measure of class, and that is the terminology of the presentation here. Fielding (1995) uses "social class" in the British sense to examine how second-generation immigrants do in the United Kingdom.
4. The ratio is similar to one used by the Federal Interagency Forum for Child and Family Statistics in its annual report, America's Children: Key National Indicators of Child Well-Being, 2002 (*www.childstats.gov*).
5. An alternative to using a market-basket measure, or an income range, is to use an income threshold to measure the middle class, though obviously a threshold will include incomes in the upper range of the distribution as well. For example, Rodriguez (1996) in his study of Hispanics chose the median income for the total population as a threshold. In contrast to the studies of thresholds and ranges, Reed (1999) uses a ratio of the income at the 75th percentile to the income at the 25th percentile. Although she was primarily interested in income inequality, the 75th/25th ratio is also a measure of the middle 50% of the income distribution.
6. Most working definitions suggest some form of economic middle class, but even an economic definition can range widely from a specific income to the income needed to buy a particular combination of goods and services. Clearly, a set income range captures part of what it is to be middle class, but in fact it is the ability to own a house, buy a car, have health insurance, and pay for college education that is the way in which the income is translated into a middle-

class lifestyle. However, in the end most definitions depend heavily on an income classification. Even those who have emphasized a market basket of goods have tended to translate that basket of goods into an income range.

7. Levy (1998) further restricted it for families in the prime earning years of 25–54, and although we will examine all household heads 18 years and older we will also examine two finer age breakdowns of the data.
8. The substantive analysis of the middle class is based on household incomes. This is an appropriate unit of analysis for our study of the middle class, as it is households, or families, with or without children, who are central in the progress of immigrant households. Such households often pool resources and move upward by a concerted effort of all households members. It is true that the results would be slightly more conservative if we used individual head of household incomes.
9. Hochschild (1995) also argues that there are flaws in the American Dream, especially for African Americans. She notes that not everyone was able to participate equally and that perhaps the resources may no longer balance the dreams. In such cases effort and talent may not guarantee success. But regardless of the flaws, the dream is obviously alive and well as the dreams of doing better are a fundamental element of the continuing pull of America.
10. It would be ironic, of course, if the American Dream were to work for the new immigrants but failed for the native-born African American population. But as in the case of the immigrant populations, many African Americans have been able to move up to join the middle class. Lingering barriers to mobility is a problem of the underclass.
11. This 1998 book by Lionel Sosa is designed to show Latinos how to market themselves to a wider American business culture. It is an unusual case study of how to "achieve the American Dream" by using both the Latino heritage and the successful practices of American business.
12. I am indebted to Peter Morrison and Calvin Beale for this anecdote that illuminates the enduring nature of the immigration process. See also Loewen (1988).
13. In general I use the terms "illegal" and "undocumented" interchangeably, though the latter is increasingly the term of choice to describe the foreign born who have entered the United States without inspection. There are of course a limited number of immigrants who may be in an undocumented and potentially illegal status for reasons other than unauthorized entry, but these are a small number.
14. Proposition 187 had the unforeseen outcome of increasing the likelihood of naturalization. Congressional decisions to cut benefits to noncitizens naturally stimulated legally admitted immigrants to become citizens.

CHAPTER 2

★ ★ ★

Immigrants in the United States

Numbers, Flows, and Policies

The size and composition of the immigrant population in the United States has changed remarkably in the past three decades.[1] In 1970 the foreign-born population was not a large presence—less than 10 million persons and a little more than 4% of the total U.S. population. By 2000, foreign-born persons in the United States were more than 28 million, and they have increased since then, making up about 11% of the total population in the United States. A large proportion (two of every five) are recent arrivals, having entered the United States within the past decade. The consensus among most demographers is that the foreign-born population will be the major contributor to the continuing growth of the nation's population. Thus an in-depth analysis of the progress of the immigrant population requires a general review the dimensions of the foreign-born population as we enter the 21st century.[2]

Historically, the U.S. population has always included a sizable foreign-born component. At the beginning of the 20th century, at the time of the last great waves of international migration, although total numbers of the foreign born were smaller, the proportion was larger (Figure 2.1). In fact, the proportion of the foreign born declined between the turn of the

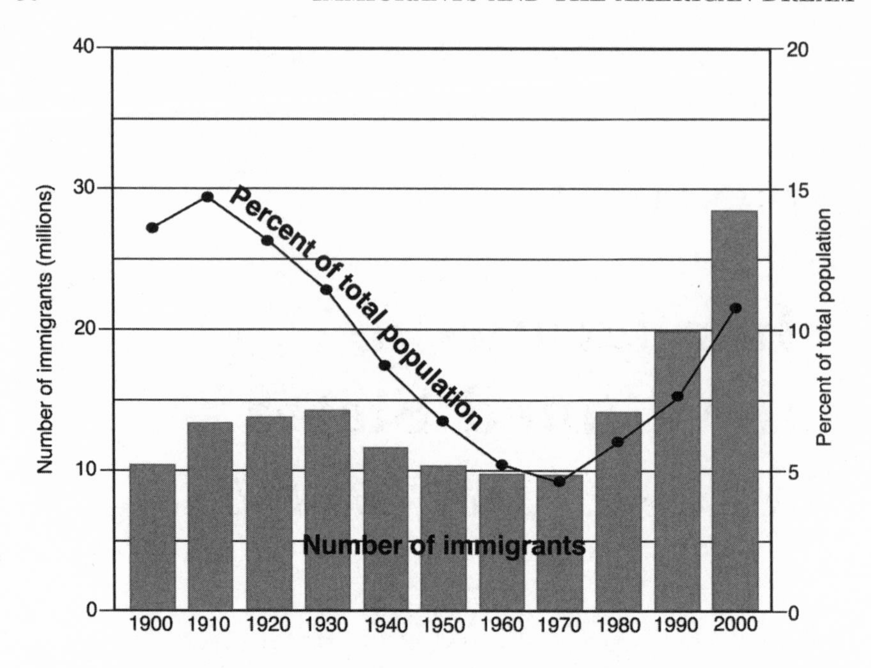

FIGURE 2.1. Number of foreign-born persons in the United States, 1990–2000. *Source:* U.S. Immigration and Naturalization Service, 1999, and U.S. Bureau of the Census, Current Population Survey, 2000.

last century and 1960 as the native-born U.S. population expanded with the impacts of the baby boom. Recently, over the past two decades of the 20th century, the growth rate of the foreign-born population was greater than at any time in the past. During the last great waves of migration, before and after the beginning of the 20th century the growth of the foreign-born population averaged about 30% a decade. In contrast, the native-born population was growing much more slowly, at about 15% per decade. During the last three decades of the 20th century the growth rate of the foreign-born population has been even higher—more than 40% per decade—while the native-born population has been growing more slowly than it did 80 years ago.

The influx of the foreign-born population and further derivative growth through births to the foreign-born population contribute most of the population growth in the contemporary United States. Census Bureau estimates suggest that births to the foreign-born population account for about half of all the growth in the U.S. population. That is, one-tenth of the U.S. population contributes about half the ongoing growth in the population, largely because the foreign-born population is more youthful and has more children than the native-born U.S. population. Moreover, whereas many immigrants once returned home after a brief sojourn of months or years, contemporary immigrants more often stay on, settling permanently rather than returning to their homelands.

Recent Census projections suggest that the U.S. population may grow to somewhere between 400 and 550 million people in the next 50 years, depending on the changes in the fertility of the foreign-born population and the level of immigration to the United States.[3] It is this context that a view of the current foreign-born population provides a broad context before I examine the central question of the book: how they are doing in achieving the American Dream.[4]

PROFILE OF THE FOREIGN BORN

The immigrant influx over the past 30 years has added 23.8 million foreign-born persons to the nation's population. The nearly 29 million foreign-born persons living in the United States in 2000 is the largest ever in the nation's history. Many are of prime working age—between 25 and 45 years old—but there are many new immigrants who are elderly family members too. Nearly two-fifths of the foreign-born population has arrived within the past decade.

Altogether 60% of the foreign-born population is settled in just the four biggest immigrant destinations: California, New York, Florida, and Texas. California alone is home to a quarter of all new immigrants and eclipses the other states, both in the total number of the foreign born and the almost 3 million recent arrivals. But immigration is not confined to just a handful of states. Ten states have more than half a million foreign born residents (Table 2.1). While about 10% of the U.S. population is foreign born, the proportion in the largest immigrant states is much higher. The proportion who are recent arrivals is perhaps the most striking feature of Table 2.1. For most states nearly 40% of the foreign born has arrived in the last decade, and in some states, like Michigan, where immigration is a relatively recent phenomenon, the proportion is nearly half.

Another useful strategy to portray the impact of the growing foreign-born population is to examine the percent of the population who speak a language other than English (Figure 2.2). The graph reflects the relative concentration of the foreign born in the very large immigrant states, but it also captures some smaller states where there is a new and large immi-

TABLE 2.1. Foreign Born (in Thousands) in 2000 by State with at Least 500,000 Foreign-born Residents

State	Population	Foreign born	Percent	Arrived 1990–2000	Percent
California	33,958	8,781	25.8	2,876	32.8
New York	18,508	3,634	19.6	1,338	36.8
Florida	15,052	2,768	18.4	1,143	41.3
Texas	20,049	2,443	12.2	1,055	43.2
New Jersey	8,105	1,208	14.9	443	40.4
Illinois	12,166	1,155	9.5	467	36.7
Massachusetts	6,194	769	12.4	327	42.5
Arizona	4,874	630	12.9	254	40.3
Virginia	6,842	526	7.7	240	45.6
Michigan	10,127	518	5.1	248	47.9
Total US	274,087	28,379	10.4	11,206	39.5

Source: U.S. Bureau of the Census, Current Population Survey, 2000.

grant population. We do not always think of Rhode Island or Connecticut as high-immigrant-impact states, and indeed the total numbers are much smaller than in California, New York, Florida, and Texas. At the same time, nearly 20% of the population in Rhode Island speaks a language other than English at home.

The 28 million foreign-born persons in the United States are a complex mixture of ethnic origins, places of birth and socioeconomic status. That given, there are a recognized set of ethnic groupings that can be used to make a set of general observations. Slightly more than half of all the foreign born are Hispanic. Most of these are from Mexican and Central America, but there are sizable numbers from the Caribbean and South America generally. Nearly, one-third of all immigrants are from Mexico and Central America, the nearest immigrant-sending regions apart from Canada. Asia as a whole has contributed about one-quarter of all the foreign born, nearly 10% from China/Taiwan/Hong Kong and the Philippines. European foreign-born persons also continue to be a significant proportion of all the foreign born (Figure 2.3). Apart from Mexico, no single country contributes much more than 2 or 3% of the foreign-born population. The ranking of sending countries clearly establishes the dominance of Mexico as the major contributor to the foreign-born population in the United States, but the wide range of countries that have contributed some immigrants includes almost every country in the United Nations. Several countries or combinations of countries have contributed

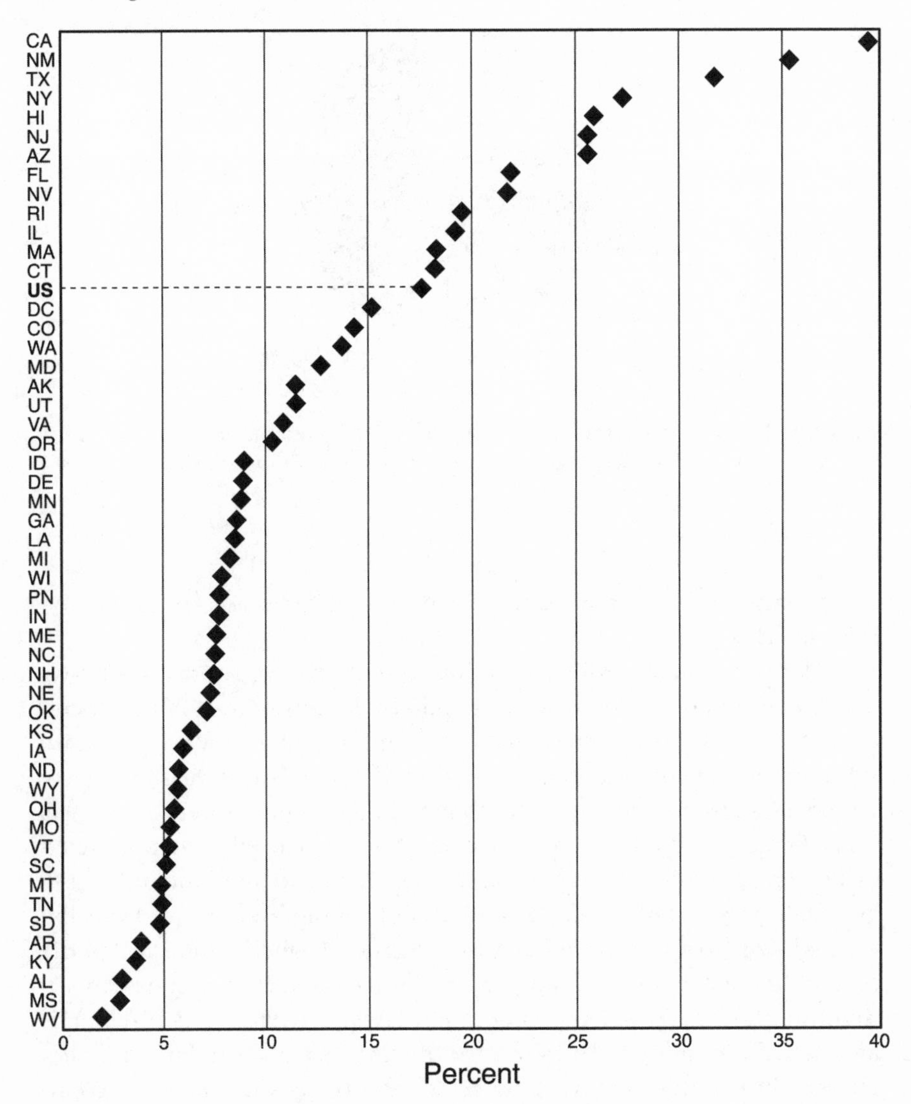

FIGURE 2.2. Percentage of the U.S. population speaking a language other than English at home (population 5 years of age and over). *Source:* U.S. Bureau of the Census, Current Population Survey, 2000, Supplementary Survey.

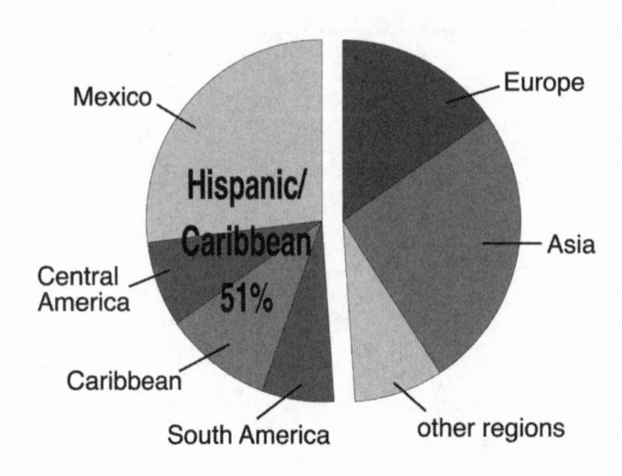

FIGURE 2.3. Foreign born in the United States by ethnic background, 2000. *Source:* U.S. Bureau of the Census, Current Population Survey, 2000.

more than a million immigrants to the present foreign-born population (Figure 2.4).

The dominance of Mexico is a function of proximity, previous flows, and the all-important network of familial relationships—a network created in the past three decades that carries information about opportunity from immigrants back to those who would seek to better their lives elsewhere. It is also a function of the change in the immigration law in 1965 that altered the quota basis of immigration to one which emphasized skilled labor and family reunification. The change to family reunification and skill-based quotas shifted the basis of immigration away from the earlier flows from Europe to flows from Mexico, Central America, and the several Asian nations that had previous links to the United States. In addition to the change in the immigration laws, the 1960s and 1970s initiated significant refugee flows from Southeast Asia and the former Soviet Union. These flows from wartime allies and from a wide variety of countries in Eastern Europe created a proliferation of the foreign-born population.

Much has been made of the change in the flows and proportions of immigrants before and after the 1965 change in immigration law; without those changes it is likely that the current diversity in the United States would not be nearly as great. The current ethnic background of the foreign born can be nearly directly related to that change in the immigration law. The diversity of recent ethnic origins includes a wider set of cultural and linguistic backgrounds than was true of the flows at the turn of the last century. At the same time the *perception* of the flows in 1900 was that of a very diverse set of origins, although of course it was less diverse than the flows today. While the immigrants from 1900 to the end of the great waves of immigration in 1924/25 were mostly from broadly European backgrounds and were often Catholic, Protestant, or Jewish, the current composition of the foreign born is a composition of Middle Eastern Muslims, Indian Sikhs and Hindus, Korean and Chinese Buddhists, and Mexican Catholics. Thus, the ethnic origins have changed, the religious backgrounds have grown more varied, and the national origins have become much more diverse than at any previous time and perhaps even more diverse than in any one nation now or in the past. The "melting pot," in short, holds a more varied and complex mixture of ingredients, not altogether like those of earlier eras.

The complexity of that mixture is reflected in the graph that reports

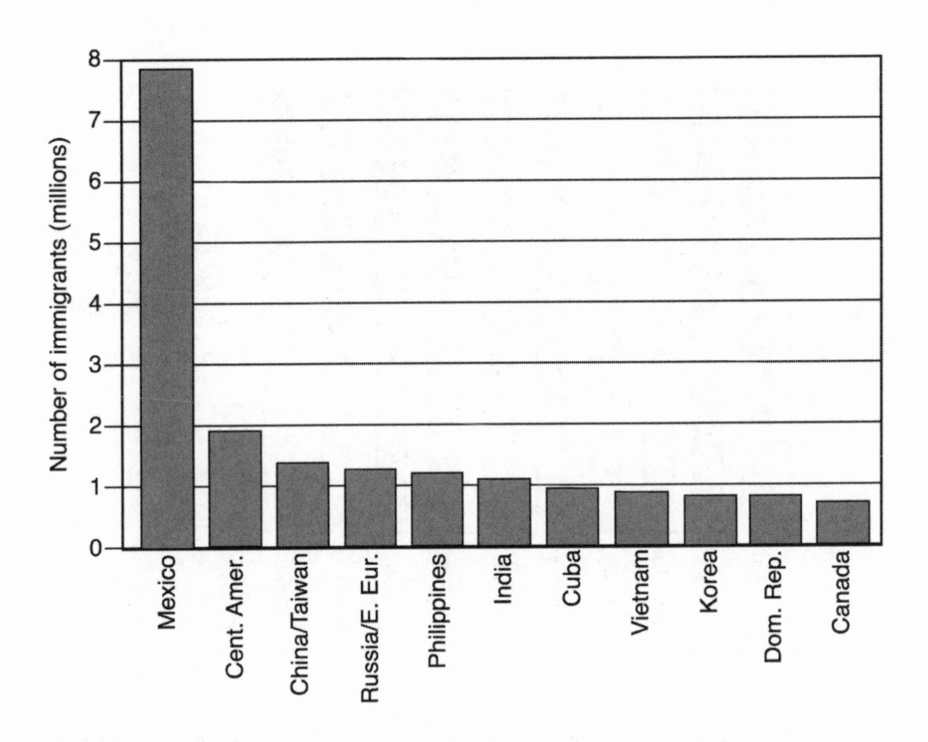

FIGURE 2.4. Major origin of the foreign-born populations in the United States. *Source:* U.S. Bureau of the Census, Current Population Survey, 2000.

the proportions of selected groups who arrived in the last two decades of the 20th century—i.e., the recent ingredients joining the mixture (Figure 2.5). One-third to one-half of all the large groups by country of origin arrived in the last 20 years. For most countries, the flows in the last decade of the century increased, in some cases dramatically.[5] For example, 25% of the Indian foreign-born population arrived in the 1980s; some 50% of the Indian foreign born arrived in the decade between 1990 and 2000. Given the nature of family reunification and the pressure to bring in more skilled technology workers, we can expect these flows to continue in the coming decades. Cubans and Canadians, by contrast, mostly arrived earlier. Overall, the patterns of large recent arrivals reflects the fundamental change that has been occurring in U.S. immigration patterns.

The foreign-born population is large and increasing rapidly, and the newcomers are also more dispersed than were earlier waves of immigrants. The immigrants are still concentrated in New York and California, though there is now a tendency to select more varied locations than was true of previous waves of immigrants.

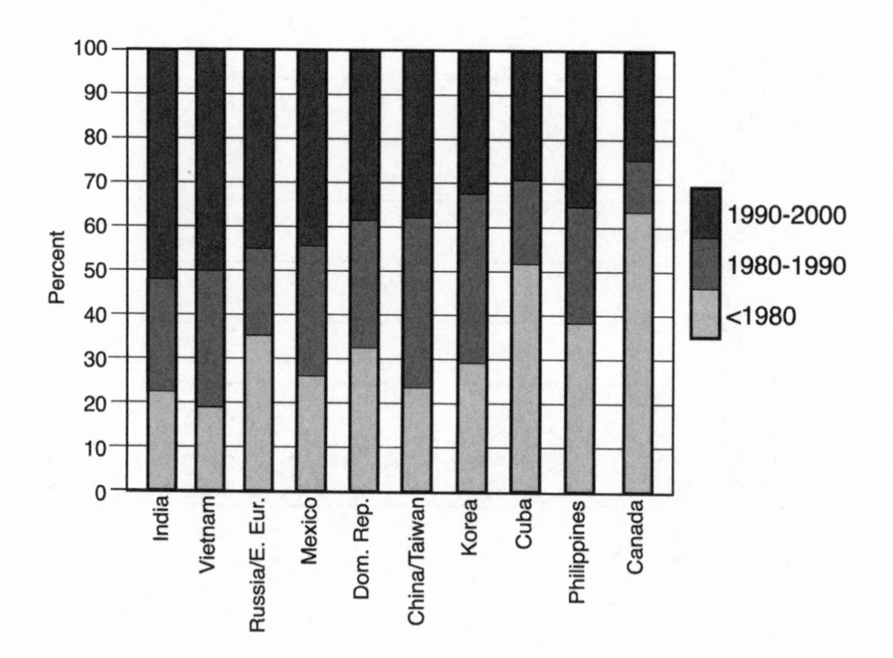

FIGURE 2.5. Proportion of foreign born by origin who entered the United States in the last two decades of the 20th century. *Source:* U.S. Bureau of the Census, Current Population Survey, 2000.

WHERE THE FOREIGN BORN LIVE

California remains the nation's major immigrant entry port: 31% of all foreign-born persons live in that state. At the same time, many other states now have noteworthy proportions of foreign-born persons (Table 2.2). Half of the states have close to 5% foreign born, and the redistribution of the foreign-born population is likely to continue rapidly in the coming decades. The proportion of the foreign-born population is high in the major states of concentration, but a few states have very high relative proportions even though their total number of immigrants is not large. Nevada and Hawaii both have moderate-sized foreign-born populations, but the proportions are 15 and 16%, respectively, nearly as large as that of Florida.

TABLE 2.2. Percentage of Foreign Born by State in 2000 (Ranked)

State	%	State	%
California	25.9	North Carolina	4.4
New York	19.6	Alaska	4.2
Florida	18.4	Iowa	3.9
Hawaii	16.1	New Hampshire	3.8
Nevada	15.2	Nebraska	3.7
New Jersey	14.9	Wisconsin	3.6
Arizona	12.9	Vermont	3.5
Massachusetts	12.4	Oklahoma	3.2
Texas	12.2	Missouri	3.0
Colorado	9.8	Pennsylvania	2.9
Illinois	9.5	Louisiana	2.6
Maryland	9.0	Ohio	2.5
Connecticut	8.8	Indiana	2.4
Oregon	7.8	Kentucky	2.4
Rhode Island	7.8	Maine	2.2
Virginia	7.7	Arkansas	1.8
Washington	7.4	Tennessee	1.8
New Mexico	5.8	Alabama	1.6
Kansas	5.7	South Carolina	1.6
Utah	5.5	North Dakota	1.5
Idaho	5.3	South Dakota	1.4
Michigan	5.1	Wyoming	1.0
Minnesota	5.1	Mississippi	0.9
Delaware	4.7	West Virginia	0.9
Georgia	4.4	Montana	0.8

Source: U.S. Bureau of the Census, Current Population Survey, 2000.

Spatially, the distribution of immigrants is still coastal and in the West.[6] The foreign-born population is concentrated along the Atlantic seaboard, in Florida, on the Gulf coast, across the Southwest, and on the Pacific coast (Figure 2.6). In general, states in the Midwest have relatively low proportions of foreign born (Figure 2.6); exceptions are Illinois and Minnesota (mainly a concentration in Chicago and Minneapolis–St. Paul). While the West has nearly 40% of all the foreign born, the Midwest, in contrast, has a little more than 10% of those born abroad. It is also true that, as in the past, the immigrant concentrations are often highly localized in older big cities, though this is changing rapidly as opportunities in small towns, particularly in retail and service activities, are increasingly taken up by immigrant entrepreneurs.

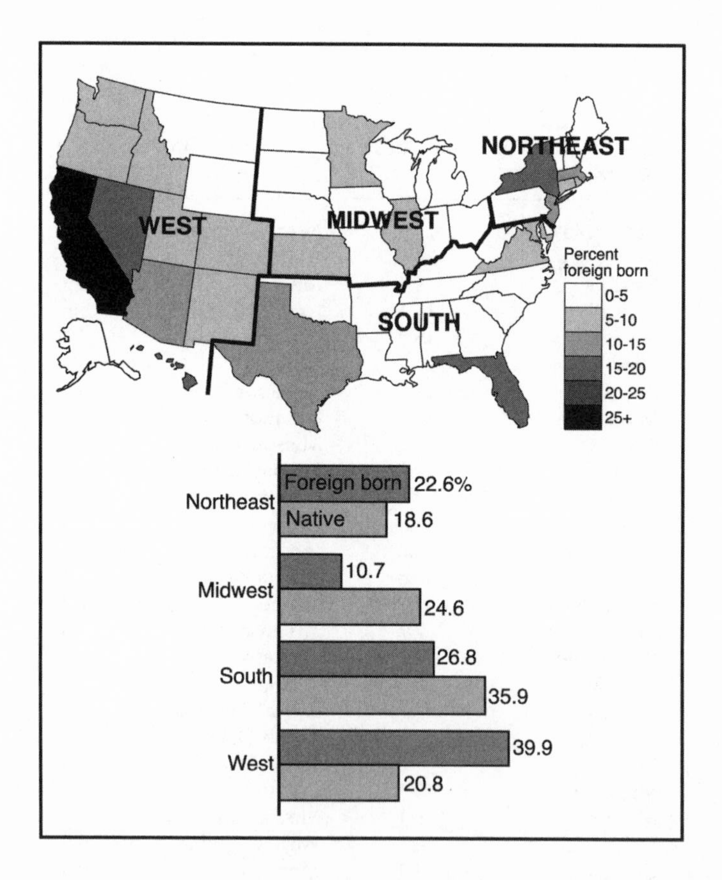

FIGURE 2.6. Distribution of the foreign born by states. *Source:* U.S. Bureau of the Census, Current Population Survey, 2000.

Although it is changing, metropolitan immigrant concentrations are still higher than for the states as a whole. The Los Angeles, San Francisco, New York, Miami, and Chicago metropolitan areas have approximately 14 million immigrants, nearly 50% of all immigrants in the United States. Nearly 43% of the Miami metro population is foreign born, and 30% of the Los Angeles population is foreign born.[7] Much of the population growth of these metropolitan areas is directly related to the increase in the foreign-born population and especially the Hispanic foreign-born population (Suro and Singer, 2002). In some areas—Los Angeles, for example—the growth in the immigrant population offsets population loss as native-born inhabitants move away or die.

Immigrants move largely to states where there are already sizable foreign-born populations, but we are beginning to see flows to many smaller states, and Southern and Midwestern states as well. Immigration is now affecting all the states and regions of the United States. Where once California was seen as different—an outpost of new immigrant concentrations, somehow different from the rest of the country—that is no longer true. Recent data from the 2000 Census emphasize the spatial changes which are underway across the American landscape (Figure 2.7). While the largest numbers of the foreign born are still in California, New York/New Jersey, Florida, and Texas, the big gains are in a wide variety of states. There are very large

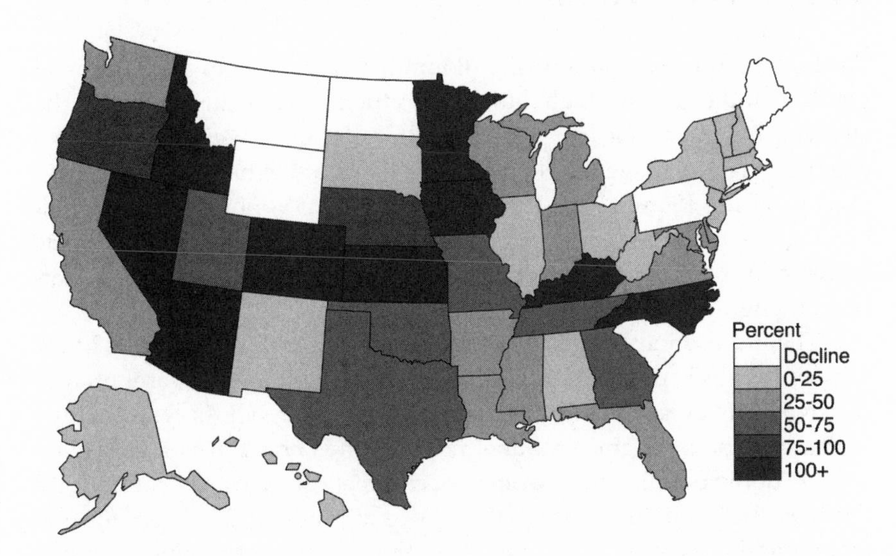

FIGURE 2.7. Percent change in the foreign born in the United States from 1990 to 2000. *Source:* U.S. Bureau of the Census, Current Population Survey, 2000.

percentage gains in many of the Western states, including Arizona, Nevada, Idaho and Colorado. But there are big percentage gains in Iowa and Minnesota. There are states that have had declines in the foreign-born population, mainly in some of the Northwestern mountain states and in Maine, South Carolina, and Connecticut. The losses in Pennsylvania suggest that immigrants too are leaving large metropolitan areas where jobs are scarce. That a broad band of states across the center of the county experienced such a large gain in the last decade is a reflection of the changing distribution of the foreign-born population. We know, too, that once the new patterns are in place further increases in the foreign born are likely. The changes once identified with a few states, especially their large cities, are more and more an apt description of changes in the country as a whole.

Even where absolute numbers are still only modest, growth rates are often dramatic, especially in the medium- and smaller-sized cities in the South, the Midwest, and the West. While most of the foreign born are still located in large cities (78% in cities of a million or more), 16% are in smaller cities and nearly 6% are in nonmetropolitan areas (U.S. Bureau of the Census, Current Population Survey, 2001; Lollock, 2001). Clearly new patterns are emerging across the nation.

AGE DISTRIBUTIONS AND FAMILY GROWTH

The foreign-born population, like all immigrant populations, is in general youthful and prone to a high rate of reproduction by virtue of being in the childbearing years. Nearly 44% of the foreign born were in the prime childbearing years, ages 25–44, while only 29% of the native born were in this age group (Figure 2.8). The predominance of young adults of childbearing age among immigrants is even greater when we reflect that the native-born population includes ethnic groups with continuing high fertility, though this will decrease slowly.

The implications for fertility and the continuing impact of the Mexican and Central American population is illustrated in the subset of age–sex pyramids by population size and age (Figure 2.9). It is clear that the increase in the immigrant resident population is being driven largely by the size of the population of women of childbearing ages. Even as fertility declines, the population born to the new immigrants will increase for several decades. The graphs for the other large immigrant groups are much more similar to those for the native born. They have larger numbers of older immigrants, though they, like the pyramids for Mexicans, have few very young foreign-born persons.

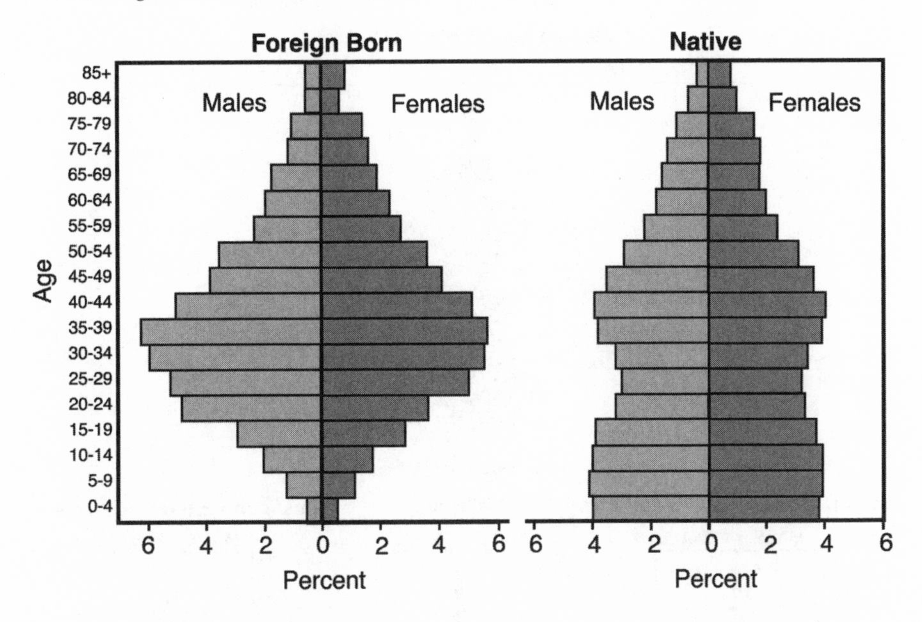

FIGURE 2.8. Population of the United States by nativity, age, and sex, 2000. *Source:* U.S. Bureau of the Census, Current Population Survey, 2000.

The "pyramid" for the Latino foreign-born population is more properly described as a diamond than a pyramid. The graph begins its bulge in the 15- to 19-year-old age group but swells considerably in the 20- to 40-year-old age categories. It also tapers away rapidly after age 54 and is somewhat weighted toward male foreign-born Latinos.[8] The age–sex pyramids for Chinese, Philippines, and Russian–Eastern European foreign born are much more balanced across the ages, although each exhibits certain distinctive features reflecting different periods of entry, and the jobs and linkages that brought them to the United States. The larger numbers, relatively, of older migrants from Asia and Eastern Europe also reflects the refugee policies which allowed these migrants entry to the United States. While the number of Mexicans who are older is presently small, that group will increase as the Mexican-origin population "ages in place."

Overall, foreign-born households are larger on average than native-born households because of higher fertility and the tendency for families to double up in a household. This is a particular outcome for migrants from Central America (Figure 2.10). Household size for immigrants from Latin America in general is nearly three times that of the native-born population. For Central America and Mexico it is more than three times larger. By using a measure of the percentage of households with five or

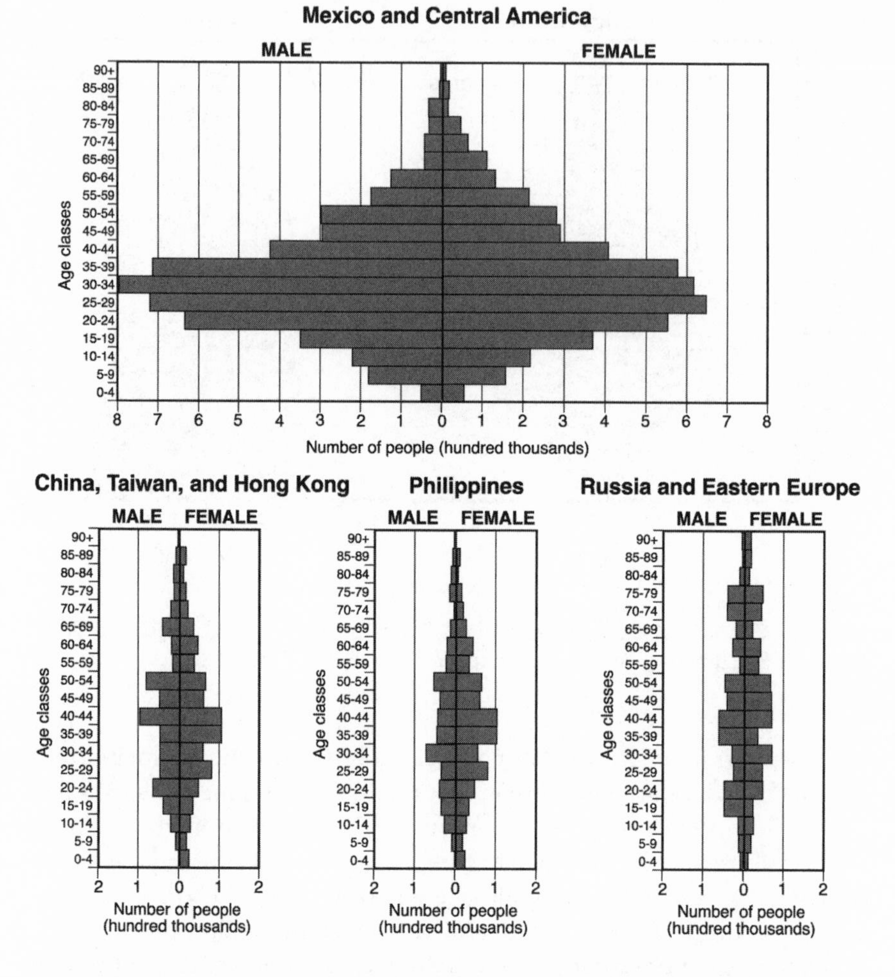

FIGURE 2.9. Age–sex pyramids for immigrants from Mexico and Central America; China, Taiwan, and Hong Kong; the Philippines; and Russia and Eastern Europe. *Source:* U.S. Bureau of the Census, Current Population Survey, 2000.

more persons, the graphs capture both the likelihood of larger families and the tendency to have more than one family in a household.

The implications for communities with larger proportions of large families shows up in the pattern of school-age children. Nearly half of the school-age children in California are either foreign born or the children of foreign-born parents (Figure 2.11). These percentages are high not just for California but for other states with high immigrant populations as well,

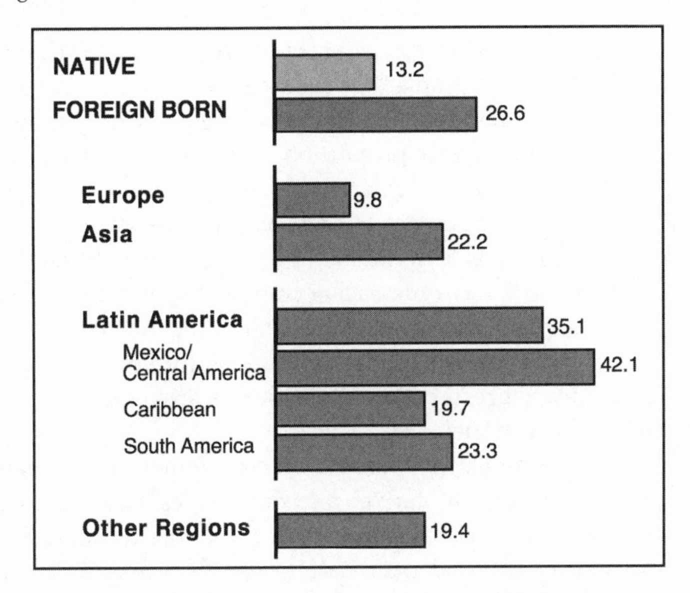

FIGURE 2.10. Percentage of family households in the United Sates with five or more members by nativity and world region of birth, 2000. *Source:* U.S. Bureau of the Census, Current Population Survey, 2000.

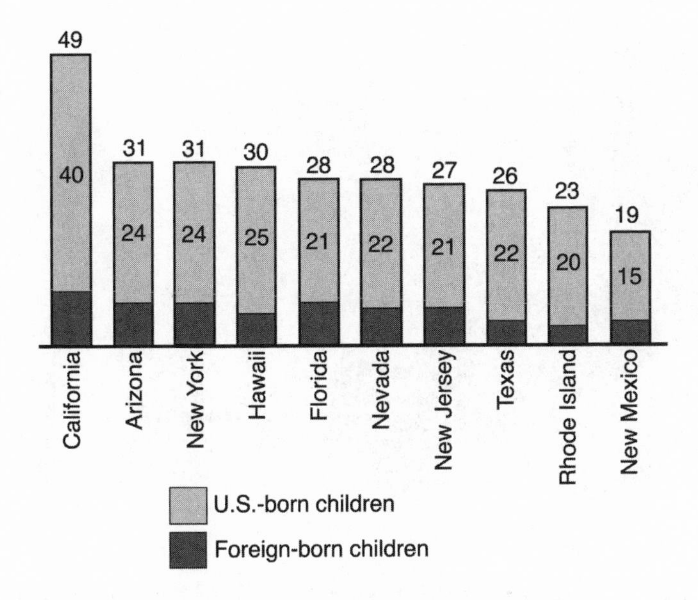

FIGURE 2.11. Percentage of children whose parent(s) are foreign born by state. *Source:* Urban Institute, 2000.

and sometimes for states that have not traditionally been the destinations of large numbers of immigrants. For example, in Rhode Island, one-fifth of the school age population are children of foreign born parents.

It is true that the largest proportion of children with foreign-born parents are still in the main immigrant gateways, but the proportions of such households are changing in states throughout the West (Figure 2.12). The Midwestern states, once dominated by school-age children in native-born households, now have increasing numbers of children of foreign-born parents. The proportions are noticeable in the Northeast and some middle-Atlantic states as well. The implications for education are obvious: at the very least, greater teaching resources are needed for students with limited English proficiency.

Particular cities similarly exemplify the prominence of foreign-born residents and the issues that derive from their educational needs. For instance, Des Moines, Iowa, illustrates how the rapid increase in the foreign-born population and their native-born or foreign-born children are transforming urban school systems. In Des Moines, the number of immigrant students in the school system has tripled since 1990, but the number of English-as-a-second-language teachers has been expanded by

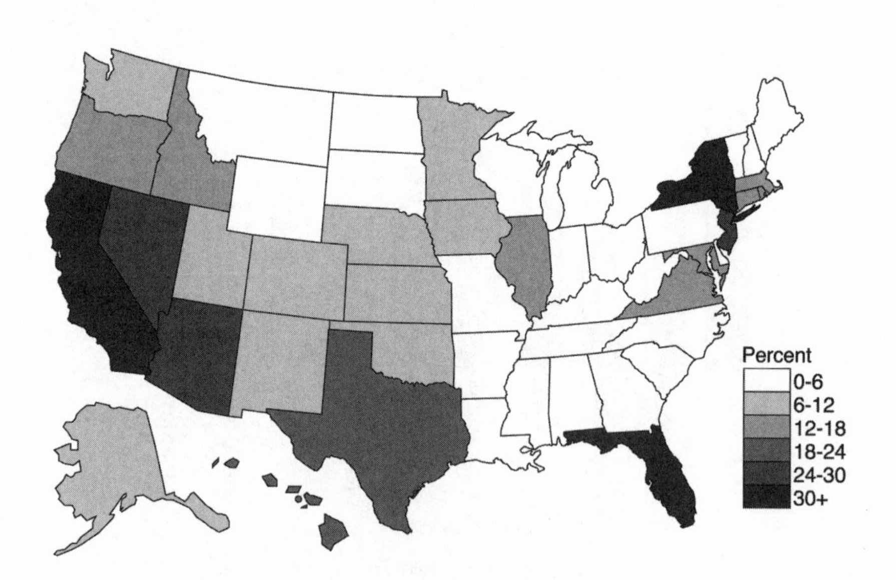

FIGURE 2.12. Percentage of U.S. households with school-age children with at least one foreign-born parent. *Source:* U.S. Bureau of the Census, Current Population Survey, 2000.

only one-third. Schools which were traditionally white, or white and African American, are now multiracial and have many different languages and cultures. Similarly, the Dallas school system is now more than 70% Latino, with associated needs for bilingual teachers and extra help for students with limited English proficiency (Clark, 1995). In Lowell, Massachusetts, Cambodian and Laotian refugees are nearly a tenth of the population, and nearly 40% of the city is foreign born. More problematic is the fact that nearly 53% of the Asian-language households are identified as linguistically isolated. A "linguistically isolated household" is defined as one in which no person of age 14 or older speaks only English and no person who speaks a language other than English also speaks English very well.

The population changes are not just changing the composition of the U.S. population as a whole, they are fundamentally changing the cities and towns of the United States. The questions which naturally arise focus on how they will blend in and whether they will be able to move up the socioeconomic ladder. It is useful to reiterate an observation that is often overlooked in discussions of blending and social integration. In such diverse settings, the new immigrants have to learn to adjust not only to local residents but to other immigrants as well.

More communities than ever now find themselves engaged in an absorption process as new immigrant streams branch out from a few immigrant gateway cities. At the same time we know that immigrant flows generate spatially concentrated migration systems, with highly focused sources of in-migration and/or sources of highly focused out-migration (Morrison, 2000). Chain migration and channelization create particular concentrations of particular groups in particular places.[9] The Hmong (Laotian Montagnards) in Fresno, California, and now in the Fox River Valley in Wisconsin, provide only one notable example of these processes at work in communities in the United States. The outcomes of these processes are seen in the changing demographics of local school systems where the children of immigrants enroll.

It is in the schools that the process of upward mobility begins. It has always been the schools that have, with greater or lesser success, provided the courses which create human capital and professional skills for individuals, and in turn created the educated labor force the United States needs. It is still in the schools that the dreams can be turned into reality. At the same time, the school systems are now dealing with a very large proportion of the children of the foreign born and the future success of the children of these immigrants is heavily dependent on the adequacy of the school systems.

SOCIOECONOMIC STATUS AND HUMAN CAPITAL

The extent to which the new immigrants and their children can and do access the educational opportunities in the United States will have far-reaching consequences for the future of the U.S. economy. Human capital is the key to the future success of the foreign-born population, as I noted earlier, and the outcomes of gains in human capital will be a central measure of progress into the middle class. Here I briefly examine the educational status of the foreign-born population and draw some conclusions for future paths in the economy.

Are immigrants less educated? Yes. Are immigrants more highly educated? Yes. Not unexpectedly, the proportion of the population age 25 and over who had completed high school was lower among the foreign-born population (Figure 2.13), a simple outcome of many immigrants who arrive with limited education from their home countries.[10] Yet, ironically, a higher proportion of immigrants also hold graduate and professional degrees. Overall, the educational backgrounds of the arriving foreign born are more complex than for those who arrived in the last great waves of immigration at the beginning of the 20th century. At that time,

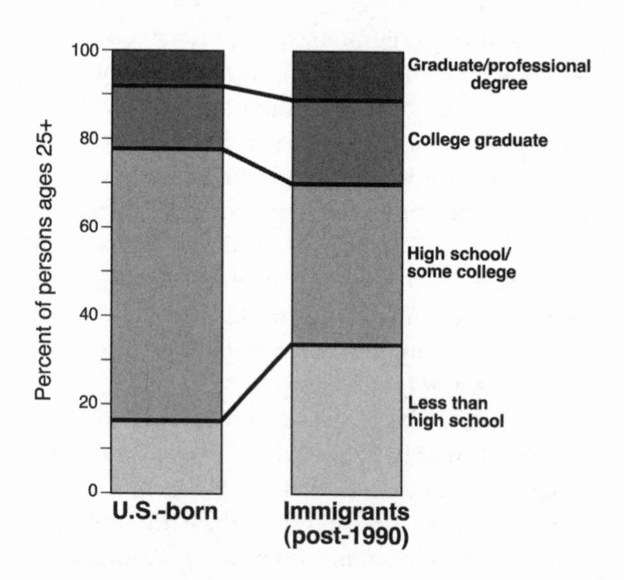

FIGURE 2.13. Contrasting educational patterns for the native born and immigrants in the United States. *Source:* U.S. Bureau of the Census, Current Population Survey, 2000.

advanced education for both the native-born and non-native-born population was a privilege for only a few, and in general the native-born population was more alike than different from the foreign-born population.

The major contrast is between the native born and foreign born with less than a high school education. Whereas the native-born population overall were more likely to be a high school graduate or have some college education (Figure 2.14), the foreign born were four times less likely to be so. However, at the college-degree level there was no difference between the foreign-born and the native-born population. This is a totally new aspect of recent immigration and fundamentally changes the mix of immigrants. Thus, the findings from the Current Population Survey for 2000 confirm other studies that the foreign-born population in the United States is bifurcated into a relatively skilled, well-educated subset and a relatively unskilled, uneducated subset (Clark, 1998). The bifurcation is being maintained by the recent inflows, which are slightly less likely to have a high school education and slightly more likely to have a college degree (Table 2.3). More than a third of the recent arrivals have not completed high school, but again nearly a quarter have a college degree or a graduate degree.

However, one should note marked differences by immigrant origin. Using the measure of at least a high school education, the rates vary from 37 to 87% (Figure 2.15). Nearly the entire native-born population has at least a high school education, as do immigrants from Europe, Asia, and

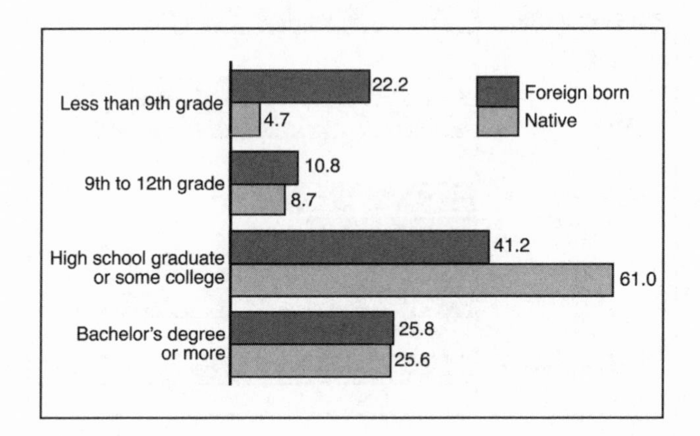

FIGURE 2.14. Population of the United States by nativity and educational attainment, 2000. *Source:* U.S. Bureau of the Census, Current Population Survey, 2000.

TABLE 2.3. Educational Attainment by Immigrant Status

Education level	All foreign born	Arrived 1990–2000	Native born
Less than high school	32.7	36.4	8.0
High school	27.3	26.1	33.4
Some college	16.9	14.7	30.0
College degree	14.7	15.4	19.4
Graduate/professional degree	8.4	7.4	9.3

Source: U.S. Bureau of the Census, Current Population Survey, 2000.

"other regions," but not many from Latin American: the gap between the immigrants and native born is greatest for those from Mexico and Central America, where only about a third of the immigrants have at least a high school education. The education gap has huge implications for the future employment trajectory of the new immigrants. At the same time, the proliferation of relatively low-skilled service jobs has provided continuing employment opportunities for the substantial numbers of Central American immigrants who have followed other family members into Los Angeles and other metropolitan areas along the Mexican–U.S. border. Those

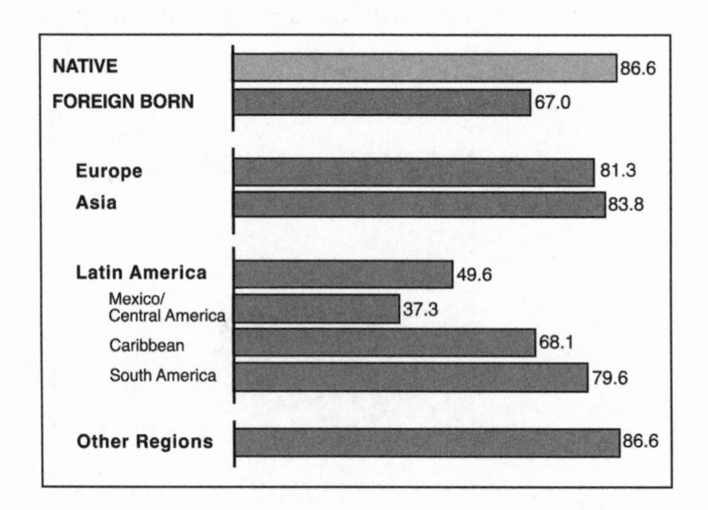

FIGURE 2.15. Population of the United States with at least a high school education by nativity and region of birth, 2000. *Source:* U.S. Bureau of the Census, Current Population Survey, 2000.

flows continue to fill the low-skilled jobs in the garment industries, in janitorial and gardening businesses, and in countless jobs that have fueled the decade-long economic expansion in the United States.

The low-skilled jobs pay poorly, and on average the foreign born earn less than the native born. More than a third of foreign-born full-time year-round workers earned less than $20,000 annually. The corresponding figure for native-born workers was 21.3%. However, a reexamination of earnings by origin shows considerable variation (Figure 2.16), indicating that there are wide differences across foreign-born groups and that many are doing much better than the U.S. median. While immigrants from Mexico, Cuba, and the Dominican Republic are below the U.S. median income level, immigrants from India are earning nearly twice the U.S. median and those from Korea, China, Canada, and even former refugees from Vietnam are outperforming that level. Clearly, many of the foreign born are competing successfully in the U.S. economy.

It is useful to reflect on the earnings outcomes for the most recently arrived foreign-born persons and households. For the most part, recent immigrants are not faring as well as immigrants overall, an expected out-

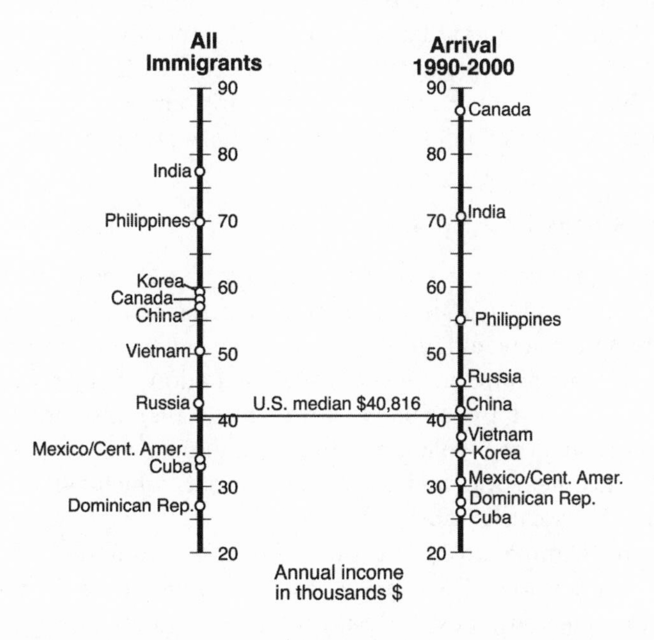

FIGURE 2.16. Immigrant median household earnings in the United States. *Source:* U.S. Bureau of the Census, Current Population Survey, 2000.

come. In most cases the median earnings are several magnitudes smaller for those immigrants who arrived after 1990. Indeed, as shown in later chapters, the decline in earnings by period of arrival is a consistent finding across most immigrant groups and over the past three decades. At the same time, skilled immigrants from India, Canada (whose recent immigrants are earning higher median incomes than Canadians as a whole), and the Philippines continue to earn more than the U.S. median. Recent Vietnamese and Chinese immigrants are close to the national median but have significantly lower incomes than their group as a whole.

The most remarkable finding may be the relative incomes of the lowest-skilled migrants. Median household incomes for the Mexican/Central American foreign born is almost $34,000, and even the most recently arrived immigrants are earning average household incomes of more than $30,000. Even though many immigrants are struggling, the median incomes suggest just how well many new immigrants are actually doing, including the most recently arrived. It is true that medians mask the distribution of incomes, a subject discussed in a later chapter. But is clear from even these median incomes (recalling, though it may not be necessary to do so, that half are *above* the median) that many are doing well and are significantly above the poverty level. While immigrants from the Dominican Republic are struggling at the bottom of the income scale and are earning only about half of the U.S. median, they are clearly earning vastly more than they could in their homeland. As the income scale suggests, the potential for the foreign born is significant.

Relative Gains and Losses

How can we reconcile the rather positive interpretations of the previous pages with the extensive discussion of the increasing poverty and lack of advancement for new immigrants? An extensive literature documents the numbers of immigrants in poverty, the stagnation of their wages, and their failure to "catch up" to native-born whites. Yet it is clear that at least some groups are doing what immigrants did in the past, steadily advancing up the income ladder and putting down roots which will lead to success in the American economy.

How immigrants do and the process of incorporation varies across a number of dimensions. The concept of modes of incorporation is one useful way to understand how the paths of incorporation vary for different immigrants. As outlined by Portes and Zhou (1992, 1993), incorporation is influenced by the polices of the host government, the values and prejudices of the receiving society, and the characteristics of the ethnic community which receives them (Portes and Zhou, 1993, p. 83). More specifically, incorporation is going to be influenced by the resources available though networks in the ethnic community in which they arrive: is there support, moral and material, which helps new arriving immigrants? It is also influenced by whether or not there are niches of opportunity which may not need advanced education—working in a small ethnic business, for example. Perhaps most important, how new immigrants do will be influenced by the conditions at the time of their arrival, both economic and political. Of course, the recent flows of relatively highly educated immigrants, and their success, reiterates the other critical variable in the modes and rates of incorporation—the resources that immigrants bring with them.

Two factors are particularly important in influencing rates of incorporation: one relates to the timing of immigrant entry; the other relates to the composition of the immigrant population. I will explore both briefly after discussing the nature of the conflict about immigrant trajectories.

There is no question that immigrants in general have been losing ground *relatively;* that is to say, the wages of immigrants in proportion to native-born wages are lower today than they were two or three decades ago. That gap between the native born and the most recently arrived foreign-born population has been increasing decade by decade (Borjas, 1998; Clark, 1998). The most recent arrivals are doing less well than earlier arrivals did. Some of this can be explained by a decline in the skill levels of recent immigrants, certainly the relative if not the absolute skill levels. Educational attainment was about one and a half years less than the native born in 1970 and was two and a half years less than the native born 30 years later (Clark, 1998). This relative decline has been in part created by the steadily increasing proportion of adult Americans with some college education or a college degree. The U.S. population is more and more likely to be well educated, whereas the educational profile of many new immigrants still resembles that of earlier arrivals. These statistics have generated an intense debate about the future of immigrant success, the impact of immigrants on the American economy, and the question of whether contemporary immigrants will ever "catch up" to the native born.

There are two quite different perspectives on how things will work out for the new immigrants. According to one view the relative skills of immigrants have not only declined over time but their assimilation is slower than in the past and earnings are likely to stay below the earnings of the native born for several decades (Borjas, 1998). In contrast, the findings of the National Academy of Sciences (NAS) study of new immigrants are much more sanguine (Smith and Edmonston, 1997). While acknowledging the problems faced by new immigrants and the tensions between earlier and later arrivals in competition for jobs, the NAS study suggested that "eventually" the immigrants will be assimilated into both the American society and the American economy. Other studies that have used cross-sectional comparisons have also emphasized the positive gains for immigrants. Several studies concluded that immigrant socioeconomic status improves with time in the United States (e.g., Chiswick, 1978; Fix and Passel, 1994). The critique of these more optimistic studies is that the gains of previous waves of immigrants is related to their higher skill levels at entry and that the new waves are less skilled than previously. Even this conclusion is debated by those who point out that in fact education levels have not declined absolutely but only relative to the gains of the native born (Simon and Akburi, 1995).

The debate about the economic gains of immigrants goes to the heart of their future social trajectories here. Those who are concerned over decreasing skill levels, and an increasing gap between the immigrants and the native born, would move to a skill-based quota for immigration. Those who see the old paths of economic assimilation being open and available find the new flows a central part of the continuing vitality of the American economy. Taking jobs at the bottom was as much a part of the past immigrant experiences as becoming a middle-class entrepreneur is today.

Clearly, both these stories are embedded in the continuing arrival of the foreign born and are central to understanding the debates about skill levels and wages—debates about immigrant success and achievement versus ethnic poverty and struggle. The United States, historically and in the present, has always been a place of opportunity, although at first there is inequality as migrants struggle to move up the social and economic ladder (Lieberson and Waters,1988). It is the acquisition of human capital, which is arguably as important as eliminating discrimination, that will slowly, if painfully, integrate society and overcome patterns of inequality. At the same time, there are two additional stories, one about the composition of the immigrant flows and another about the most recent waves of immigrants, which help explain what is happening.

First, aggregating all immigrants masks the successes and failures of different immigrant groups. Second, the very recent large-scale flow of less-skilled immigrants has reduced the overall average levels of skills and earnings. Large-scale flows of relatively unskilled immigrants will reduce mean skill levels and wages of all immigrants, but this does not mean that skilled immigrants are doing less well nor that they are unable to make the upward shifts in earnings, homeownership, and professionalization. Yet to the extent that very large numbers of new immigrants arrived in the 1990s, there will be, at least in the short term, a greater aggregate negative impact on skills and earnings. When many of these new immigrants are undocumented and are in unstable and temporary work environments, it is not surprising that there has been relative slippage for immigrants in the aggregate.

POLICIES, IMMIGRANT FLOWS, AND GLOBAL CONNECTIONS

The number of the foreign born in the United States and the annual flows are not independent of the way in which the United States as a nation has viewed immigration over time nor are they unrelated to global economic changes.

The outcomes of the changing policies, or the lack of policies, is easily seen in a time line of the changing number of immigrants admitted to the United States (Figure 2.17). The juxtaposition of flows in the first de-

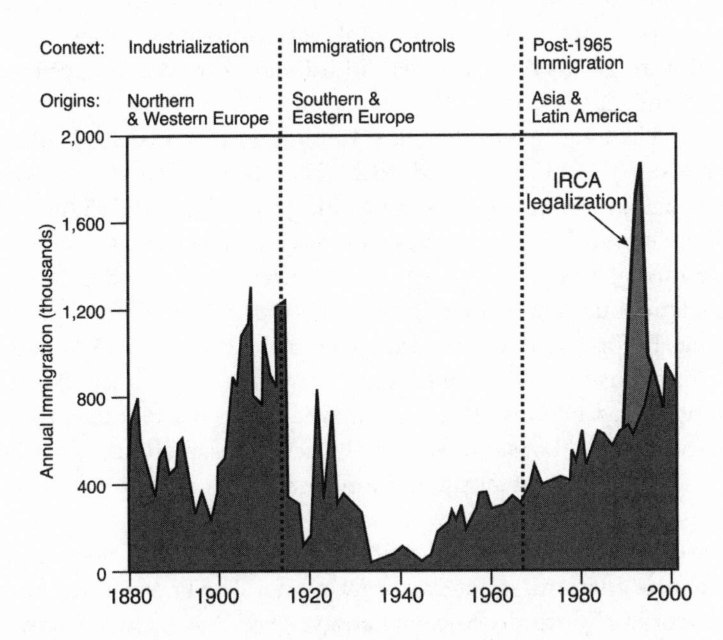

FIGURE 2.17. The historical context of U.S. immigration. *Source:* U.S. Immigration and Naturalization Service, 1999.

cade of the 20th century and the last decade of the 20th century has been noted frequently, but at the same time the steady increase in immigrant arrivals beginning in the 1950s often goes unremarked. The number of immigrants arriving in the United States has been growing quite regularly; it is not a simple and sudden onrush of new immigrants. Apart from the IRCA (Immigration and Reform Act of 1986) legalization program, the flows have been growing steadily. What has changed and what is the main concern of this book is the composition of those immigrants. The Hart–Cellar Act in 1965 changed the preference system to emphasize migration based on skills and on family reunification rather than on national quotas. The latter had privileged immigrants from Western Europe. The change allowed many more immigrants from Asian and Latin American nations to apply for visas to the United States.

Two other important policy changes also impacted immigration to the United States. First, as immigration numbers increased and a backlog of applications under the family-sponsored immigration quotas increased, pressure mounted to raise the numerical limits and to legalize those immigrants already in the country but without documentation. The IRCA of 1986 was designed to deal with the backlog of applications, make those unauthorized immigrants who were in the United States legal, and provide a set of deterrents to further undocumented immigration. The legislation provided penalties for employers who knowingly hired unauthorized immigrants and provided additional resources to apprehend illegal entrants at the borders.[11]

Second, a new federal law, the Immigration Act of 1990, increased the per-country limit and added extra visas. The law went some way toward reducing the backlog and the waiting time, but by 1995 the waiting list had grown to 3.7 million applications (Weintraub et al., 1998). A further attempt at rationalizing the immigration process and dealing with the problem of undocumented immigrants was taken with the Illegal Immigration Reform and Immigrant Responsibility Act of 1996 (IIRIRA). Unlike previous acts which addressed numbers and quotas, the IIRIRA was designed as a control mechanism, oriented especially toward targeting undocumented border crossing. The act increased funding for border control and immigrant apprehension and strengthened the employer sanctions provisions of the law.

The decision to modify the welfare laws of the United States also had policy effects on immigration and immigrants. The Personal Responsibility and Work Opportunity Reconciliation Act of 1996 included provisions that removed a variety of public support from legal immigrants. The act removed Supplemental Social Security and Medicaid until citizenship was granted or the immigrant had worked a qualifying 40 quarters (in effect more than 10 years). This removal of federal support, like the decisions with respect to border enforcement, has probably had only a peripheral effect on actual numbers and flows but may well have been important in the recent increase in naturalizations.

The final dimension of changing immigration policy has been the growth in the H-1B visa program. This program is specifically focused on highly skilled "temporary" immigrants and is designed to bring in just those immigrants who are well educated and perhaps already middle class, or who are very likely to make the upward trajectories in jobs and incomes that will lead in turn to middle-class status.[12] The requirements for an H-1B visa are indeed professional training (the specialty occupations include accountant, computer analyst or programmer, engineer, financial analyst, scientist, architect, and lawyer) and college education. The visa program has been expanded in the late 1990s to nearly 200,000 per year for 1999–2002. This is a significant element of the bifurcated flow that was reviewed earlier. Even though the H1B visa program has been expanded considerably, there is greater demand by employers for visas for potential employees. At the same time the visa program has sparked a lively debate about whether it takes jobs from native-born information technology specialists.[13] For the purposes of this book it is a thread in understanding the bifurcated flow of immigrants and the increasing number of skilled and educated immigrants who are entering the United States.[14]

Global Interconnections

The migration flows that have developed in the past two decades are interconnected with the changing economies of developing and developed nations. In the past decade the economies of the developing nations, by and large, have struggled, while the economies of the industrialized countries of Europe and North America have flourished. Given the rapid transmission worldwide of images and news information on the developed economies and their demand for labor, it is no surprise that large-scale migration has been occurring globally (Castles and Miller, 1993). While it is true that it is difficult to tie direct links between international migration and global economies, there is sufficient evidence for the U.S. Department of Labor to conclude that "the timing, direction, volume, and composition of international migration . . . are fundamentally rooted in the structure and growth of the regional economy in which the United States is most actively involved. Flows of labor occur within an international division of labor with increasingly integrated production, exchange and consumption processes that extend beyond national boundaries" (U.S. Department of Labor, 1989, p. 5).

Globalization is a useful though ill-defined concept to describe the changes that are sweeping across the nation-states. The notion includes increasing international trade and investment, fundamental changes in the nature of communication (better and cheaper telephone links), greater regional integration, and increasing social interchange through visits and trips to relatives at home and abroad. There are estimates, though unsubstantiated, that one-third of the Mexican population has a relative or family member living in the United States. Despite the lack of specific definitions, it is fair to say that globalization is about the integration of people into world markets and the increasing interconnectedness of the world economy.

Capital in the global economy flows freely. Labor flows much less freely, and it is the controls on the flow of labor (the requirements for entry and residence) which in turn generate undocumented migration. Capital creates jobs, and the labor requirements, especially for low-wage labor, create immigration flows. If the legal flows are insufficient to fill jobs and if the demand for low-cost labor is substantial the structure of border control will generate illegal immigration. The undocumented immigration is a by-product of imposing controls on legal immigration. Thus, undocumented immigration is an outcome of the nature and structure of borders; well-guarded and controlled borders are seen as controls on the flows of population. Now, the economies of the world are much more interconnected than they were during the last great waves of immigration and the economies are also more segmented than in a previous time, hence the demand for both high-skilled and low-skilled workers.

OBSERVATIONS AND IMPLICATIONS

The above sketch of the basic dimensions of the immigrant population and review of the debate about how immigrants are faring can serve as a background for the detailed analysis of the trajectories of the foreign born over the past three decades. Specifically, we can ask about the complex paths that immigrants are following and investigate the way in which different groups have been successful.

The brief review in the preceding pages documents the increased intensity of immigration in the past three decades. The numbers are large, the flows fairly constant, and the proportions in many receiving states are such that the foreign-born population is rapidly approaching a fifth of the total population in the most populous states. In California, within a decade or two, the foreign-born population will likely equal the native-born population. These are profound changes.

The changes continue to be most marked in the big cities like Los Angeles, San Francisco, Miami, and New York, which are still the primary immigrant gateways. The existing immigrant settlements in these locations are the magnets for additional flows regardless of the labor opportunities in these cities. The data also show that while the distributions are still concentrated, flows are directed to states that had not been immigrant destinations. It is likely that immigrant communities serve as a "social safety net" (Reitz, 1998) for immigrants. The immigrant community provides support and security when the labor market is not as strong.

But how immigrants do is not just a function of their skills and the impact of ethnic enclaves in providing opportunities. The outcome for immigrants is also related to the context in which they arrive—the social and educational milieu that they find on arrival. It seems that some of the effects of low wages and poorer outcomes are as much related to how successfully immigrants, particularly educated immigrants, can access the system. Often, educated immigrants do end up in service-sector jobs because of the imposition of perhaps institutional gatekeeping on their ability to practice their profession or skill in the United States. It may be that U.S. institutions have imposed a burden on immigrants by making their economic transition that much more difficult (Reitz, 1998). The "warmth of the welcome,"[15] as Reitz notes, may be as important as the skills and composition of the immigrant workforce.

The composition of the flows foreshadows patterns for the future. The immigrants, as in the past, are largely youthful and prone to childbearing (though that varies a good deal by ethnic origin).[16] Even so, the very large flows of young immigrants from Mexico and Central America will have major transformation effects on the states and cities where they arrive. They have already done so in a wide variety of communities in Southern and Northern California, in Miami, and in New York City, and no doubt will do so in other large cities.

Even though many of the new immigrants are not highly skilled and are earning relatively low incomes, there is clear evidence that a significant number of the newcomers are doing well in their new country. Even immigrants who as a group have much lower levels of education, the Mexican and Central American immigrants, for example, are earning about two-thirds of the average median U.S. earnings, and many immigrants with higher levels of education are doing especially well. Clearly the levels of human capital that the foreign born bring with them, or acquire after entry, are the key to their making more rapid progress up the income ladder. It is the trajectory up that economic ladder which is the heart of the following chapters on economic and professional progress to the middle class.

NOTES

1. Varied data sets are used throughout this book and, rather than discuss each data set at different points in the text, I have included a detailed appendix on immigration data generally, and the specific data used in the analysis in this book.
2. Throughout this study we use the terms "immigrant population" and "foreign-born population" interchangeably. The foreign-born population is all those persons who are born outside the United States or its territories and who are not the children of U.S. citizens.
3. The U.S. Census makes projections based on a set of assumptions about fertility and immigration. Those projections have a low, middle, and high range. Current projections suggest that the middle projections will increase the U.S. population to 403 million by 2050 and 570 million by 2100. The higher projections suggest that the population may reach 809 million by 2050 and 1.2 billion by 2100. The outcome will depend on how migration and fertility change in the coming century.
4. Much of this material is also well reviewed in other publications (Clark, 1998; Camarota, 2001).
5. For China, Taiwan and Korea the flows have been somewhat even in the two decades 1980–1990 and 1990–2000. Only for the Philippines has the *rate of flow* decreased substantially over the two decades.
6. See McHugh (1989) for the distribution a decade earlier.
7. The data are for Consolidated Metropolitan Areas.
8. There is some interesting switching back and forth between male and female age groups. For example, females between ages 35–44 dominate the foreign-born groups from China and the Philippines. Both pyramids from China and the Philippines are somewhat more skewed to female immigrants, especially the Philippine distribution, whereas the Russian and Eastern European pyramid is more balanced, though again there is considerable shifting across gender and age groups.

9. Chain migration and channelization (earlier arrivals creating continuing links to homeland towns) are the processes whereby immigrants establish links between places. Earlier arrivals are followed by later arrivals who are thus "channeled" to particular geographic locations.
10. In general, Mexico does not provide education beyond elementary school. Thus many of the immigrants who arrive from Mexico and by extension from other Central American countries will have only limited education, especially if they are from rural areas where schooling is even more problematic.
11. There is substantial evidence that the IRCA did not slow the flow of undocumented immigrants to the United States (Bean, Chapa, Berg, and Soward, 1994).
12. The H-1B visa program is not a program that entitles the visa holders to permanent resident status, but in effect many of those who enter on H-1B visas are able to change their status to that of permanent residents.
13. Hearings about extending the number of H-1B visas can be found online at *www.ins.usdoj.gov*, and the debates are summarized in N. Matloff's testimony to the U.S. House Judiciary Committee at *http://heather.cs.ucdavis.edu/itaa.html*.
14. The H-1B visa program is closer to the Canadian skills-based admission program.
15. The words quoted constitute the title of the book by Jeffrey Reitz (1998).
16. The proportion of the growth that will come from continuing high fertility is open to considerable debate. Already there is evidence that the fertility of the foreign-born population is declining, and almost certainly it will continue to do so. However, births per 1000 foreign-born Hispanic women aged 15–44 were nearly twice that of native-born women. And even if the large number of foreign-born women of child bearing age have only replacement fertility of two children for each adult woman, there will be significant effects on the growth of the U.S. population. A detailed discussion of recent immigrant fertility can be found in Bachu and O'Connell (2001).

CHAPTER 3

★ ★ ★

Making It in America

The Foreign-Born Middle Class

American society has always emphasized the possibility of upward mobility, personal improvement, and financial success.[1] Where homeownership was once a distant dream, it is now the cornerstone of a lifestyle that includes for many people at least one, usually two cars, adequate health insurance, a retirement plan, and income sufficient to provide a college education for children and even grandchildren. Clearly, the middle-class lifestyle is a central part of the iconography of the United States and carries its images well beyond the shores of America. How does this dream intersect with the flow of new immigrants and the increasing numbers of the foreign born? Are they too a part of this success story? Are they also becoming middle class, with all its connotations of material success and socioeconomic status?

This chapter focuses on the overall success of the foreign born. It is followed by more detailed discussions of professional gains, entry to homeownership and political participation—what we can term the icons of middle-class status. For the most part, the discussion is about the foreign-born middle class, but some data on ethnic native-born and white middle-class households is included for comparative purposes. There is also a growing African American middle class, but that story is beyond the scope of this book.

The idea of immigrant progress was bolstered by past research which suggested that, even though immigrant earnings were lower when the immigrants first arrived, in a few decades they caught up to the native-born population. The research even suggested that immigrants not only caught up but that for the same skill levels they surpassed the earnings of native-born Americans. These ideas at first sight are consistent with our ideas of immigrant arrivals in earlier time periods. Immigrants came with few skills but were able, by hard work and the translation of the skills that they did have, to move up the earnings ladder. However, this sanguine view has been countered by more recent analyses which show that the successive waves of immigrants to the United States are doing *less well, relatively,* over time. That research emphasizes that immigrants, especially the latest arrivals, are losing ground in comparison with the gains of the native born (Borjas, 1998). These debates, however, are based on analyses of average wage and participation rates, without considering what is happening to particular groups of immigrants. Indeed, it may be harder for the most recent immigrants to make the gains that were available to earlier arrivals, but lost in these debates is the question of whether there is a growing segment of the immigrant population that is doing well.

The issue of how many immigrants are successful, who among them are, and why is a central part of current efforts to come to grips with how American society will reinvent itself with a new and diverse immigrant population. If there is a large emerging immigrant middle class, it speaks to the success of the American Dream. The debate about immigrant success in total is also important because it highlights the question of the future trajectory of the foreign born. And, to the extent that it asks about social integration into the middle class, it is asking the most important question about economic success, the question of economic integration into the larger U.S. society.

But the debate is not unrelated to the general economic progress of American society as a whole. Since the 1980s there has been a growing concern about increasing income inequality and the "vanishing middle class." Thus, the questions about the trajectories of new immigrants are questions that are being asked also about the native-born population, especially the question as to whether the previous paths of upward mobility and greater success are being changed. I consider such questions further in the succeeding chapters. However, for the foreign born, the questions about the future of the middle class are questions about the ability to be a part of what has always been seen as quintessentially American—the ability to rise with merit from humble beginnings.

DEFINING THE MIDDLE CLASS

In the preceding chapters, the definition of the middle class was based on a combination of income *and* homeownership.[2] The two measures are not unrelated, but together they capture both the income which is generated by the household and the wealth effect of homeownership. Homeownership can be viewed as a proxy for a wealth effect; indeed, ownership is a very large part of the wealth of many "middle-class" households.[3] Moreover, to the extent that ownership is closely tied to other assets, and in turn to retirement plans, the combination of income *and* ownership goes a long way toward an adequate measure of middle-class status. Because the American Dream implies the promise of increasing incomes over time and the advance into homeownership, the measure here is a good surrogate for measuring progress toward its fulfillment.[4]

The previous discussions of social integration suggested that for immigrants the process of moving up is also the process of economic assimilation. While the assimilation process is multidimensioned, certainly much of the process depends on success in the economic realm, the process of becoming structurally similar to the host population. In economic terms, full integration would mean that the occupational and income distributions would be very similar across all ethnic and racial groups. It is asking too much to expect such similarity within only a decade or two, but the extent to which the process of economic advancement is set in motion will lead eventually to greater assimilation.

For the native born and the foreign born alike, the changes in the economy have emphasized the importance of education specifically and human capital more generally. Education and, by extension, skills are critical to the process of making economic gains in the "new economy." In a context of societal economic change, time of entry also takes on an important role in the process of social integration. The process of moving up the economic ladder is influenced by the context of entry, since immigrants who enter during a period of economic expansion are certainly going to have an easier time in moving up than immigrants who arrive in times of economic recession. It is true that some recent immigrants are entering the United States and the middle class simultaneously, but in general new immigrants move up the economic ladder slowly and the second generation has a higher probability of being middle class than first.

Other recent studies, specifically those focusing on the progress of Latinos, have also suggested that newcomers are doing what newcomers have always done: slowly, often painfully, but quite assuredly embracing the cultural norms of the United States (Rodriguez, 1996, 1999). Based on an analysis of citizenship, homeownership, language acquisition, and intermarriage, Rodriguez concludes that immigrants are assimilating today in the same way that immigrants did in the past.[5] The biggest worry is whether progress can be sustained (E. Lopez, Ramirez, and Rochin, 1999). There are real concerns about educational gains in particular. It seems that only about 10% of third-generation Latinos have university degrees (bachelors or more advanced degrees) whereas 30% of the native-born non-Latinos are at this education level. Yet the central issue is about whether this is just happening more slowly for some groups than for others and whether in the long run all groups will follow the paths carved out by earlier immigrants to the United States.

The remainder of this chapter uses income and homeownership in combination to define the middle class. To reiterate, the argument emphasizes that middle-class status is a combination of both income level and housing status. It captures the notion that both the ability to buy the middle-class lifestyle and the commitment to and integration into the local community, represented by ownership are essential parts of middle-class status. Of course, in this analysis as in other research on the middle class, the numbers and proportions that are discussed in the book are a direct function of the definitions, and different definitions will indeed provide different estimates of the middle class. However, the similarity of the combination measure of income and homeownership to other "middle-income ranges," as Chapter 1 documented, provides support for the approach in this book.

BECOMING MIDDLE CLASS

The foreign-born middle class is a complex mix of recently arrived households who are already middle class and those who have worked their way up after arriving as young children. Still others are the native-born children of earlier immigrants who are now successful members of the ethnic middle class; this third group has made the classic generational move up the social and economic ladder. They are the successful children of immigrants who arrived two or three decades ago and who worked at menial jobs to ensure that their children would be successful.

At the beginning of 2000 there were, by my definition, approximately 31 million middle-class households in the United States, nearly 30% of all U.S. households. There were 2.7 million foreign-born middle-class households, 20% of all foreign-born households, and they made up 8.6% of all middle-class households. If we add in middle-class households of ethnic native-born populations, Asian and Hispanic, on the premise that most of them are the children of immigrants, the immigrant and second-generation middle-class total is 3.8 million. It is more than 12% of all middle-class households in the United States. The media perceptions of a growing ethnic middle-class population are thus consistent with the statistical data.

The number of middle-class households is large and growing (Table 3.1). The Hispanic and Asian foreign-born middle class nearly tripled in the two decades from 1980 to 2000. The white foreign born had a smaller growth rate, though as a group it still added more than 100,000 new middle-class households. Native-born ethnic groups grew as well, and made up about 400,000 new middle-class households, but the growth of the native-born ethnic middle-class households is dwarfed by the growth of the foreign-born numbers (Figure 3.1). Equally notable is the finding that the increase in the foreign-born middle class plus the native-born Hispanic

TABLE 3.1. Middle-Class Households in the United States by Immigrant Status and Ethnicity

	1980	1990	2000	Percent increase 1980–2000
Foreign Born				
Asian	170,660	351,462	559,240	227.7
Hispanic	294,880	539,198	882,022	216.9
White	895,460	912,281	1,004,906	12.2
Other	55,400	90,145	163,542	195.2
Total	1,416,400	1,893,086	2,609,710	84.3
Native Born				
Asian	110,660	125,199	185,941	68.0
Hispanic	635,600	647,407	919,955	50.5
White	22,669,820	24,144,880	24,546,389	8.3
Total Middle Class	26,514,560	27,699,028	30,852,744	16.4

Note. Middle class is defined as 200–499% of household poverty level income and homeownership. The total middle-class numbers include African Americans and other racial and ethnic groups.
Source: U.S. Bureau of the Census, Public Use Microdata Sample, 1980 and 1990, and Current Population Survey, 2000.

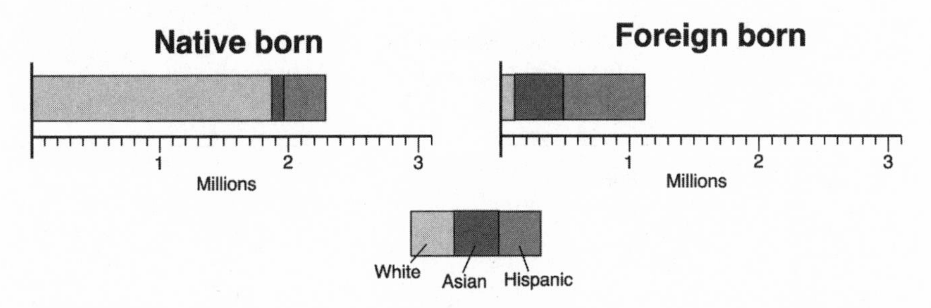

FIGURE 3.1. Growth in the middle class in the United States, 1980–2000. *Source:* U.S. Bureau of the Census, Public Use Microdata Sample, 1980, and Current Population Survey, 2000.

and Asian middle class is about the same as the increase in the native-born white and African American middle-class population.

The results are impressive—gains of more than a million middle-class Hispanic and Asian households in a 20-year period. The increase in both numbers and proportions suggests that economic prosperity, despite the downturns, is lifting the new Asian and Hispanic households, both native and foreign born.

The story of the *relative* gains to middle-class status is also positive, although it is slightly different than that about the absolute gains. Examining the proportion of foreign-born households to all middle-class households and the changes in those proportions over time provides additional insights into the progress of the foreign born. Foreign-born groups have increased their proportion of the total middle-class households, sometimes quite significantly. Overall, foreign-born households have increased from 5.3% of all middle-class households to 8.6%. The gain for Hispanic foreign-born households is from 1.1 to 3.0% and for Asian foreign-born households from 0.6 to 1.8% (Table 3.2). The total growth of native-born ethnic plus foreign-born middle-class households is from 8.1 to 12.3%—clearly a big change in just two decades.

Further insight comes from looking at the growth of foreign born middle-class households in relation to the overall growth of the middle class. Are foreign-born households reaching parity with white native-born populations? Are the increases in Asian, Hispanic and other foreign born middle-class households about the same as the increase in their populations as a whole? Examining middle-class gains in the context of overall changes in the proportion of middle-class households provides a

TABLE 3.2. Ethnic Middle-Class Households as a Proportion of All Middle-Class Households

	1980	1990	2000
Foreign Born			
Asian	0.6	1.3	1.8
Hispanic	1.1	1.9	3.0
White	3.4	3.3	3.3
Other	0.2	0.3	0.5
Total	5.3	6.8	8.6
Native Born			
Asian	0.4	0.5	0.6
Hispanic	2.4	2.3	3.1
Total	2.8	2.8	3.7

Source: U.S. Bureau of the Census, Public Use Microdata Sample, 1980 and 1990, and Current Population Survey, 2000.

somewhat more cautionary tale than that of the absolute numbers and the relative percentages of households who are middle class.

Neither native-born ethnic nor foreign-born households have reached parity with native-born white households. That is, their proportions of the population are greater than their proportions who are middle class (Table 3.3). Thus, in the year 2000, Asian foreign-born households made up 2.5% of all U.S. households but they made up 1.8% of middle-class households, a shortfall, so to speak, in their proportional representation. This was similarly true for Hispanic and foreign-born white populations. Only native-born white households outperformed their proportion of the population. Overall, there is a greater proportion of white native-born households who are middle class than the proportion of white households of the total population. In 2000, although native-born white households were only 71.6% of all households, they represented nearly 80% of all middle-class households (Table 3.3).

Graphing the difference between the proportion of households and the proportion of the middle class shows that indeed foreign-born households have not reached parity and lag behind their relative population proportion by 2 or 3% (Figure 3.2). However, hidden within these findings are some important conclusions. Foreign-born whites, native-born Asians, and native-born Hispanics were either close to parity and stable or had made progress toward parity over time (Figure 3.2). It is only the most recent foreign-born arrivals, especially Hispanics, who show a loss

TABLE 3.3. Proportions of the U.S. Population and Proportions of the Middle Class by Origin

	Asian		Hispanic		White		Other
	NB	FB	NB	FB	NB	FB	FB
1980 population	0.4	0.8	3.1	1.9	78.5	4.2	0.4
Middle class	0.4	0.6	2.4	1.1	85.5	3.4	0.3
1990 population	0.5	1.6	2.8	3.5	76.6	3.7	0.6
Middle class	0.5	1.3	2.3	1.9	83.9	3.3	0.3
2000 population	0.6	2.5	3.5	5.3	71.6	3.8	0.9
Middle class	0.6	1.8	3.1	3.0	79.6	3.3	0.5

Note. NB, Native Born; FB, Foreign Born.
Source: U.S. Bureau of the Census, Public Use Microdata Sample, 1980 and 1990 and Current Population Survey, 2000.

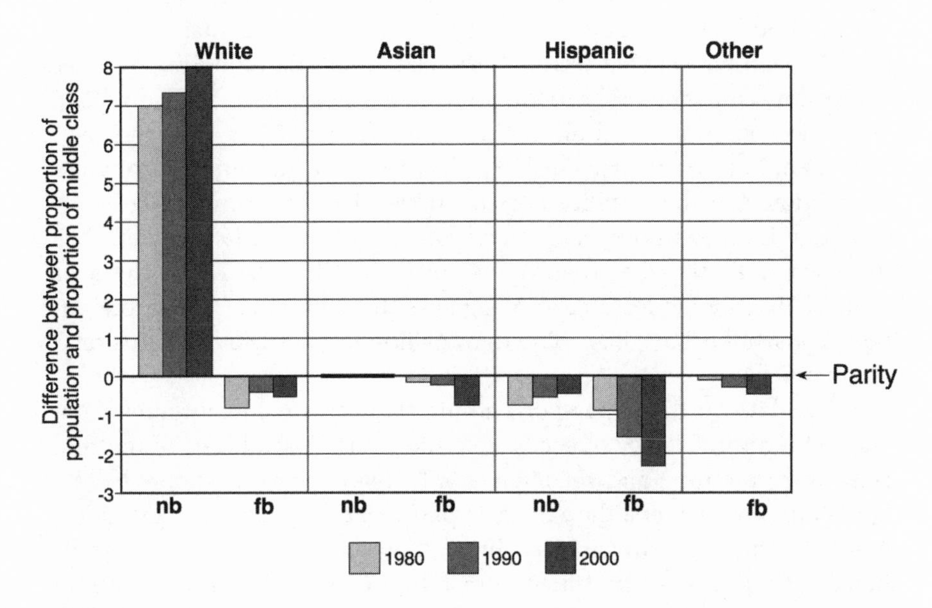

FIGURE 3.2. Gains and losses in the middle class over time in the United States. *Source:* U.S. Bureau of the Census, Public Use Microdata Sample, 1980 and 1990, and Current Population Survey, 2000.

of ground, a finding which is consistent with the earlier comments on the very large numbers of the foreign born (often unskilled) who arrived in large numbers in the good economic times of the 1990s. They are proportionately less likely to be doing as well as earlier arrivals, and their recent arrival means that they are very unlikely to have achieved middle-class status—it would be surprising were it otherwise. The very large increase in the number of the foreign born who arrived between 1990 and 2000 makes it that much harder for them to advance.

Proportionally, there has been a slight decline in all middle-class households. In 1980 about 33% of all households were middle class; in 2000 that proportion had declined to about 29.5%. The proportion of white native-born households who are middle class also showed a slight decline from nearly 36% to a fraction under 33% in 2000. The decline is consistent with the general sense that there is pressure on the middle class, but it does not support the notion of a "hollowing out" of the middle class (Figure 3.3). As for the population as a whole, there has been a small proportional decline in the total foreign-born middle class as well, but this finding is not uniform across all groups. While a smaller percentage of Asian households were middle class in 2000, the decline was reversed for white and other foreign-born households in total (Figure 3.3). Hispanic households had an initial decline from 1980 to 1990 and then stabilized. Yes, the proportions are lower than for the native-born white population, but the stability is remarkable given the continuing large-scale immigration of poorer Hispanic households in the 1990s, which has the effect of diluting the proportion who are middle class. The evidence of stability or only modest declines suggests that more than a few foreign-born households have been able to transition to middle-class status relatively quickly.

The data on the native-born ethnic changes are equally interesting and tell a positive story of socioeconomic gains. Asian levels of middle-class entry are high and relatively stable. Native-born Hispanic households are also stable; although their participation rates are about 5–7% less than those of native-born whites, they are still remarkable. Additional conclusions can be drawn from a division of middle-class status by age. Not unexpectedly, older households are more likely to be middle class—we would not expect otherwise. More interesting is the relative stability in their middle-class entry over time. Older Asian foreign-born households increase their participation levels in the middle class, Hispanic and "other" foreign-born households decline proportionately be-

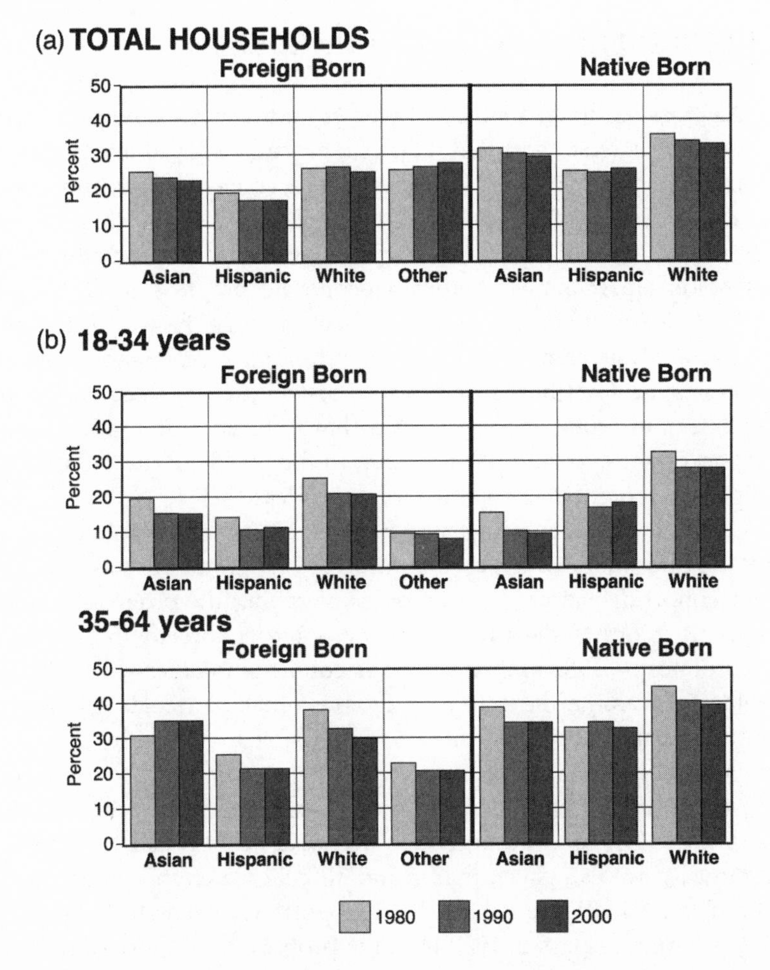

FIGURE 3.3. Middle-class households as a proportion of households: (a) total households and (b) for ages 18–34 and 35–64. *Source:* U.S. Bureau of the Census, Public Use Microdata Sample, 1980 and 1990, and Current Population Survey, 2000.

tween 1980 and 1990, and then stabilize. The fact that the proportions of younger households who are middle class is much lower reflects the very large number of less-skilled workers and households who will have to move up if the gains of past groups are to be repeated. It is quite possible that younger foreign-born Hispanic households will have a harder time making the transition than earlier arrivals did, highlighting that it may take them much longer to achieve their American Dream.

EVIDENCE FROM COHORT CHANGES

While a cross-sectional analysis of who is middle class can provide some of the evidence to suggest that there is a process of social integration and upward economic mobility, better evidence comes from examining the trajectories of age cohorts and generational changes. The cross-sectional analysis showed that the foreign-born Asian, Hispanic, and ethnic white households are making significant economic progress in certain contexts and clearly there is a real numerical increase. However, the cross-sectional analysis cannot show what individuals experience over time, progressing up (or sinking down) in a distribution of unequal incomes. To gain any insights, one needs data that trace specific groups (age cohorts) and their evolving participation in the middle class. Do they gain or slip, on average, over two or more decades?

Using cohorts also brings us closer to the earlier discussions of immigrant incorporation, which by definition is a temporal process. Following an age cohort over time is a way of asking about the progress of similar people. By plotting the progress of two age cohorts who entered the United States by 1980 and another age cohort who entered by 1990, it is possible to evaluate the extent of progress that particular groups have made over time. In this analysis it is not possible to follow the individuals in an age cohort on a year-to-year basis. But it is possible to examine *the group* of 20- to 29-year-olds in 1980 and compare them with the corresponding 30–39 age group in which they will be included 10 years later. Then in turn we can check that group 10 years later still when they are 40–49. The evaluation compares the proportion of the 20–29 age group who were middle class in 1980 with the proportion of those same households who were middle class when they were 30–39, 10 years later in 1990, and in turn in 2000 when they were 40–49.[6] Following a group is not a perfect solution to changes over time, as losses do occur to the group through mortality and emigration, but such change in the aggregate is relatively small. Thus we are able to closely approximate overall changes in the relative economic position of the cohort. Note that for the foreign born we control the cohort by examining the age group 10 years later only for those who had arrived by 1980. New arrivals are not included in the cohort. To provide a basis for comparison, we include native-born Hispanics and native-born Asians and place the foreign-born gains in the context of their gains and those of the native born as a whole.

For the United States as a whole, the foreign born are making significant movements into the middle class. Perhaps the best illustration of that progress is to focus on the 20- to 29-year-old group in 1980 who steadily increase their proportion who are middle class such that by 2000, two decades later, nearly 34% have middle-class status (Figure 3.4). They start out at lower levels than the native born do, but they reach nearly the same levels. Older households in 1980 make slower progress, and (as we would expect) at older ages they are less likely to be in the middle class. But this is not different from the native born as a group.

The 20- to 29-year-old households in 1990 made spectacular progress by 2000. Of course, these households are more nearly the "1.5 generation," as they were already here as young teenagers in the 1970s.[7] As there is a debate about the most recent immigrants and what is often described as their lack of success in the labor market, it is important to look at how the most recent arrivals are doing as well. The progress of those who arrived between 1980 and 1989 and who are measured in 2000 shows the same basic pattern of rapid progress into the middle class (Table 3.4). For all foreign born, the increase is from 6%, quite low, to nearly 19% in the middle class in a decade. The evidence is not different for the most recent arrivals.

The general patterns are still true when we break down the foreign born into Asian, Hispanic, and ethnic white groups (Figure 3.5). I have plotted the paths of the native-born ethnic groups for comparative purposes and repeated the panels which have the native-born changes.[8] While there have been questions about whether or not Asians are the "model minority,"[9] within the context of cohort progress to the middle class the term seems particularly apt. Foreign-born Asians are making

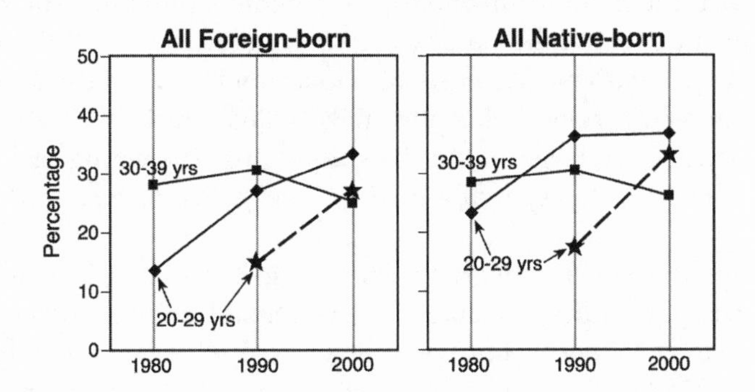

FIGURE 3.4. Changes in the proportion of middle-class households by cohort for the United States. *Source:* U.S. Bureau of the Census, Public Use Microdata Sample, 1980 and 1990, and Current Population Survey, 2000.

TABLE 3.4. Changes in the Proportion of the 20- to 29-Year-Old Middle-Class Cohort Who Arrived in the United States in 1980–1990

	US		CA		NY/NJ		TX/AZ/NM		FL	
	1990	2000	1990	2000	1990	2000	1990	2000	1990	2000
All	5.8	18.6	5.5	14.3	4.6	11.5	4.6	20.4	8.7	31.9
Asian	8.3	22.6	11.0	17.9	7.3	28.5	6.1	—[a]	10.8	35.0
Hispanic	4.0	16.4	3.6	13.1	2.1	4.2	4.3	20.6	8.1	21.0
White	8.4	21.7	7.5	13.4	7.9	20.9	6.6	51.0	11.9	58.2

Source: U.S. Bureau of the Census, Public Use Microdata Sample, 1980 and 1990, and Current Population Survey, 2000.
[a]Small sample sizes.

significant cohort progress across the past two decades. The 20- to 29-year-old households start at a proportion of 12.3% and end up at 34.2%, two decades later. They are virtually indistinguishable from the native-born population as a whole. The increase for native-born Asians is from a higher base and reaches a middle-class representation rate of 37%.

Hispanic households have greater variation in their trajectories, but the general paths of successful penetration of the middle class are still clear (Figure 3.5). It is true that foreign-born Hispanic households make much slower progress and reach only about two-thirds the level of the total native-born. The two-decade increase for native-born Hispanic households who are 20–29 in 1980 is from about 11.0 to 28.5% in 2000. The 30–39 age group have difficulty maintaining their middle-class proportions, perhaps a result of not having access to retirement programs and pension plans that support middle-class white households. A powerful finding is the result that native-born Hispanic households do almost as well as Asian native-born households and make gains which bring them both close to native-born levels of middle-class status. The youngest cohort of Hispanics in 1990 do especially well in their gains in the decade of the 1990s.

In sum, the increases overall do not bring Hispanic households to the same levels of middle-class states as the general population, but at the same time the total native-born population is also having difficulty maintaining its middle-class status. The starting levels for the native-born population are significantly higher, but while the youngest age cohort does make progress, it does not reach the levels of the same age cohorts in 1980. Again, it is useful to focus on the results for the most recent arrivals.

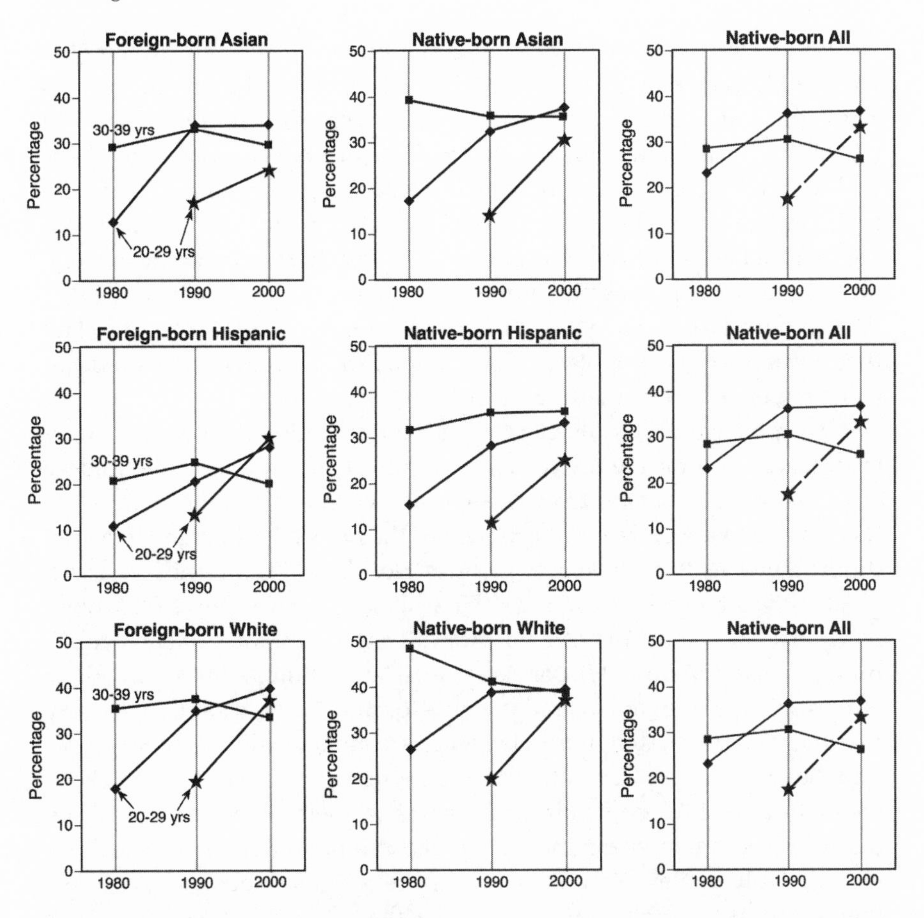

FIGURE 3.5. Changes in the proportion of middle-class households by cohort and ethnic origin for the United States. *Source:* U.S. Bureau of the Census, Public Use Microdata Sample, 1980 and 1990, and Current Population Survey, 2000.

It is true that they have very low levels of middle-class gains in 1990—only 4%—but by 2000 these most recent arrivals are 16.4% middle class (Table 3.4).

Foreign-born white households make gains in middle-class status that bring the group very close to the level of the native-born population as a whole and to almost the same levels as those of native-born white households (Figure 3.5). Households in the age cohort 20–29 who were in the United States in 1980 double their proportions with middle-class status in two decades. The older-age cohort starts high and barely maintains its middle-class status, and the youngest group in 1990 nearly doubles the level of participation. These are high levels and rapid gains in middle-class status. In addition, white households who were 20–29 in 1990 and who entered the United States between 1980 and 1990 nearly tripled their middle-class status, from 8.4 to 21.7 (Table 3.4). The gains reflect the higher human capital levels that so many ethnic-white households (from the Middle East and from Europe) brought with them when they arrived. Overall, the most pervasive finding is of convergence. While it is true that foreign-born groups have not reached the proportions of the middle class that are true of the population as a whole, they are making significant gains in being part of the middle class.

The patterns in middle-class gains vary by region (Figure 3.6). However, there are some common elements in the cohort paths to the middle class. As for the United States as a whole, Asian and foreign-born white households make more rapid progress to the middle class than Hispanics do. At the same time, Hispanic gains in Florida and California are much greater than those in New York/New Jersey or even Texas/Arizona/New Mexico.[10] New York/New Jersey has the lowest rates of entry to the middle class for both Hispanics and Asians, though rates for ethnic whites are similar to the overall patterns for middle-class gains.[11] In every case the youngest native-born Hispanics are making significant advances into the middle class. The rates for those who were here in 1980 and who were 20–29 in 1990 had quite large gains in their proportions who are middle class. Again, these gains are high for more recent arrivals too (Table 3.4). Only in New York/New Jersey are the gains substantially lower than those in the other states and lower than for the United States as a whole, and some of this is explained by the lower likelihoods of owning versus renting (which will be discussed later in the book).

In several cases, the proportion of the young foreign-born Hispanics in the middle class nearly doubles as the cohort ages 10 years. In other cases the increase is on the order of 70–80%. This is true in Florida and California, where in general foreign-born Hispanic households are pushing against the rates of entry for the population as a whole. The real contrasts are between Texas/Arizona/New Mexico, Florida, and California, on one hand, and New York/New Jersey, on the other. New York/New Jersey is an extreme case. Hispanic households in New York/New Jersey are far from the middle-class participation rates of Texas and Florida.

As for the United States as a whole, the findings by state for the foreign-born white population are a confirmation of the progress that comes with higher levels of human capital and labor market skills. Both

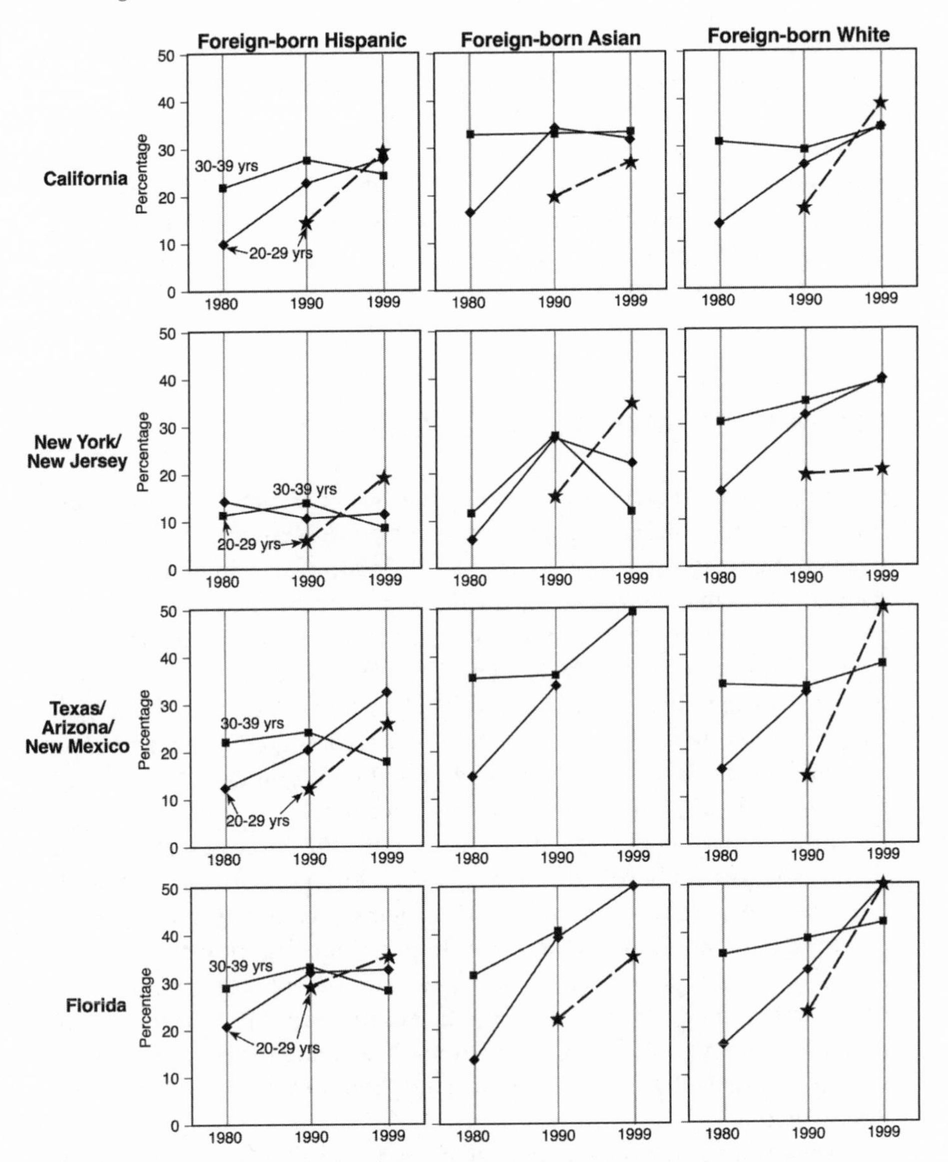

FIGURE 3.6. Changes in the proportion of middle-class households by cohort by ethnic origin and region. *Source:* U.S. Bureau of the Census, Public Use Microdata Sample, 1980 and 1990, and Current Population Survey, 2000.

cohorts made progress, and in several situations the ethnic white foreign-born cohorts reached 50% and higher in participation rates.[12] Even in New York the rates of middle-class participation by white immigrants is notable—nearly 40% for both 20–29 and 30–39 age cohorts in 1980. In some cases the data are not sufficiently rich to provide the details for older ages in the cohort survival analysis. The gains for the very youngest cohorts in 1990 are again a confirmation that the process of upward mobility is alive and well in individual locations as well as in the United States as a whole.

CLUES FROM CHANGING GENERATIONS

Beyond the cohort analysis, there are clues to what might happen in the future from how different generations succeed in becoming middle class. Even though there is not a great deal of data on the differences across generations, what exists does provide a tantalizing picture of possible paths to the future. By tracking the experiences of distinct generations, we gain hints at what may be to come. How are the second and third generations doing in their quest for middle-class status? Are they gaining or slipping on average? Where is the third generation on the status ladder? The outcomes will tell us how the children of immigrants, as they in turn become heads of households, are doing over time.

In this analysis of generations, I compare the successes of the children of the foreign born with those of the children of native-born households whose parents were immigrants, thereby contrasting the second and third generations. I examine three immigrant categories: both parents native born; one parent native born and one parent foreign born; and both parents foreign born. I examine each of these groups for two age cohorts, 30–39 and 40–49. For the categories, I calculate the proportion who were middle class based on the aggregated 1996–1999 Current Population Survey (CPS).

The results are revealing, important, and consistent with the expectations about moving up and becoming assimilated economically. The central contrast is between households whose heads have two parents who are foreign born and those whose parents are both native born. As we would expect, households with two foreign-born parents are less likely to have joined the middle class than are households with both native-born parents. That is, the data support a position of increasing likelihood of being middle class in the third generation (Table 3.5). The results are consis-

TABLE 3.5. Changes in the Relative Proportion of Second- and Third-Generation Households Who Are Middle Class

	Two parents FB	One parent FB	Two parents NB
All Asian Households			
Age 30–39	21.9	21.9	25.2
Age 40–49	30.4	27.7	40.3
All Hispanic Households			
Age 30–39	17.4	26.6	26.3
Age 40–49	23.6	32.4	35.4

Note. FB, foreign born; NB, native born.
Source: U.S. Bureau of the Census, Current Population Survey combined data file for 1996–1999.

tent for both Hispanic and Asian households. For Hispanics, there is not a great deal of difference between one or two native-born parents. The differences for Asians are more striking. Asian households with one foreign-born parent do not have quite the same proportional representation in the middle class as do households with two native-born parents. The 40–49 age cohort households for both ethnic groups have higher proportions with middle-class status than do the 30–39 age group, again as expected.

What do the generational results add to our earlier findings? They emphasize two things: that later generations are more likely to be in the middle class than their parents, and that all groups move with relative rapidity into the middle class over time.[13] The results also draw attention to the differences across ethnic groups, and by implication they point to the important role of previous education in creating the paths to middle-class status—not a new finding, but one which is worth reiterating.

WHAT DOES IT TAKE TO BECOME MIDDLE CLASS?

The previous discussions have provided a good idea of who, where, and how many of the foreign born are "making it" in the United States. Even so, it is important to extend this discussion about the numbers, proportions, and rates with a discussion of the associated factors that facilitate becoming middle class. What factors make it more likely?

To examine what affects the likelihood of becoming middle class I use simple models of the probability of being middle class as a function of the classic variables that are thought to influence upward mobility. These variables include the following: age, time in the United States, citizenship status, English language proficiency, professional status, years of education, and whether there are two workers in the household.[14] As we consider later in this chapter, when a person comes to the United States does make a difference, and the longer that person is here the more chances he or she has to move up the socioeconomic ladder. In addition, just time spent in the United States increases the chance of earning more and being in a better job. Language proficiency—being able to speak English well—is well documented for its enabling ability. Many professions require English-language proficiency, which in combination with education, is an important means of moving up the socioeconomic ladder. By definition, having a higher household income will increase the chances of being middle class; not only do two workers bring in more money to the household, they also provide the economic security enabling the family to enter the homeowner market.

The results of a simple probability model confirm these interpretations, which will be elaborated on in coming chapters on professionalization and citizenship. There are, of course, variations across the states (Table 3.6). The critical variable is whether or not there are two workers in

TABLE 3.6. Variables That Are Related to the Probability of Being Middle Class

Variable	US	CA	NY/NJ	TX/AZ/NM	FL
Age	1.00	1.00	1.00	1.01	1.00
Duration	1.03	1.04	1.03	1.04	1.03
Citizenship	1.18	1.18	1.24	1.24	1.24
Speak English well	1.23	1.50	1.14	1.31	1.09
Professional occupation	1.03	0.98	1.07	1.16	1.18
Years of education	1.03	1.03	1.03	1.06	1.05
Two workers	2.38	1.94	2.06	3.64	3.09
Percent concordant	73.7	66.5	65.3	73.5	68.9
Tau	0.13	0.11	0.09	0.15	0.15

Note. Values in the table are odds ratios. Thus, a ratio of 1.5 for speaking English well in California, for example, raises the likelihood of being in the middle class by 50%.
Source: U.S. Bureau of the Census, Current Population Survey, 2000.

the household. Thus, the foreign born are caught in the same vice as the native born: it takes two incomes to be in the middle class. For the foreign born in the United States as a whole, having two workers doubles the probability of being in the middle class. In Texas and Florida it more than triples the probability of being middle class. Several other variables are noteworthy. Time in the country (duration) is moderately important; it increases the likelihood of being middle class slightly. English proficiency substantially increases the probability of being middle class, as does U.S. citizenship. Having a professional or managerial occupation substantially increases the probability of being middle class in Florida and Texas/Arizona/New Mexico.

The probability of being middle class is much stronger for groups that have traditionally emphasized education and skill training, though it varies remarkably across groups from different origins (Table 3.7). Citizenship and professional or managerial occupations are also powerful forces across nearly all groups. However, while professional occupation is the single biggest predictor for Indian households, it is unimportant for Russian/Eastern European households. Clearly, different groups take different paths to the middle class, but for nearly all groups U.S. citizenship is inextricably bound up with progress up the socioeconomic ladder. For Mexican-origin households, U.S. citizenship is likely to increase the probability of being middle class by a third. For Korean households it is the critical factor.

TABLE 3.7. Variables That Are Related to the Probability of Being Middle Class for Selected Countries of Origin

Variable	Mexico	China	India	Korea	Russia
Age	1.01	1.00	1.02	0.95	0.93
Duration	1.03	1.03	1.08	0.77	1.02
U.S. citizenship	1.35	1.51	1.24	6.01	1.51
Professional occupation	1.66	1.34	3.08	—	0.17
Years of education	1.03	1.04	0.82	3.16	1.17
Two workers	2.76	2.79	0.93	—	11.61
Percent concordant	70.0	69.8	83.1	97.9	85.1
Tau	0.13	0.11	0.09	0.15	0.15

Note. Values in the table are odds ratios. Thus, a ratio of 3.0 for having a professional occupation for Koreans, for example, raises the likelihood of being in the middle class by three times.
Source: U.S. Bureau of the Census, Current Population Survey, 2000.

The findings from these models of who is middle class emphasize the importance of professionalization and U.S. citizenship, which are the subjects of detailed analysis in coming chapters. Even so, while occupation only modestly increases the likelihood of being middle class for some, it is useful to reiterate that the probability of being a professional is in turn correlated with speaking English well and being a citizen. Citizenship is closely involved with increasing language skills, essential in the naturalization process. In other words, U.S. citizenship, occupational status, and language proficiency are interwoven; together they impact the chances of making it into the middle class.

OUTCOMES IN REGIONS AND COMMUNITIES

Because different places provide different experiences for immigrants, examining particular places can provide information about what might happen in the future at other locations. California and New York, Miami, and Dallas–Fort-Worth can be mirrors for what is likely to happen in Iowa, Nebraska, and most of the central regions of the nation. To some extent the dispersal of immigrants has already begun, and there are media stories of new immigrant concentrations in small towns and large cities alike. Moreover, many of these stories are about middle-class immigrants, those who are making it in their new society. Clearly, there has been an increase in the proportion of middle-class households across the United States. In many states now 3–12% of middle-class households are foreign born (Figure 3.7). Nevada is notable for the high proportion of foreign-born middle-class households, as are several states in the middle-Atlantic region and the Northeast. Florida, New York/New Jersey, and California still have the largest proportion of middle class who are foreign born. Thus, the major changes and increases are still in the four big regions in which most immigrants are arriving.

California and New York have always been on the leading edge of change and nowhere more obviously than in the recent changes in immigration and foreign-born concentrations. How is California different, and is it a temporal change that will engulf other places in time? Are the rapid and dramatic changes in New York and increasingly in Florida and Texas harbingers of things to come in other regions, or are they peculiar to these states with high immigrant populations? We cannot know for sure how the changes in immigration will be played out on the large canvas of the United States, but a study of regional variations can offer some clues as to

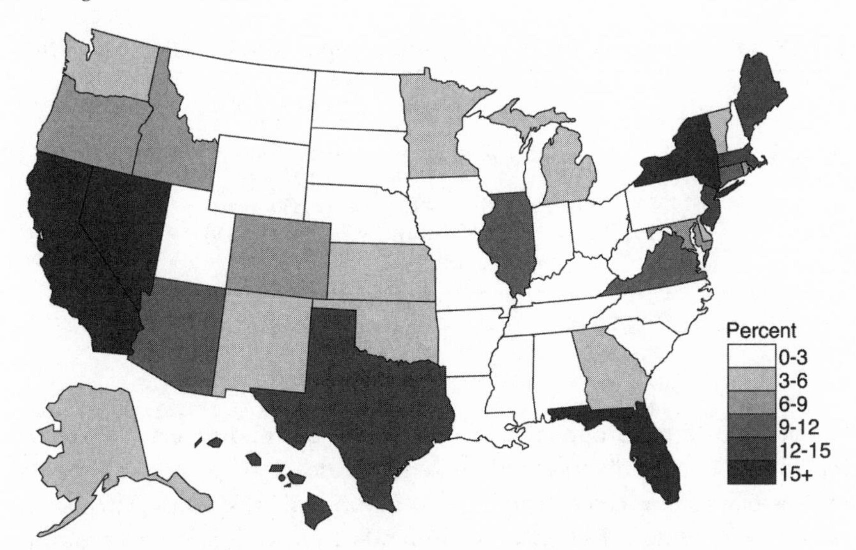

FIGURE 3.7. Percentage of U.S. middle-class households that are foreign born. *Source:* U.S. Bureau of the Census, Current Population Survey, 2000.

how the process will change and transform the immigration landscape of the United States. The following subsections examine how different immigrant groups are concentrated, their growth in the last two decades, and their current proportions of the middle class.

Numbers and Growth

California may indeed be a harbinger of the future; there are more middle-class foreign-born immigrants there than anywhere else in the nation. In California, the total numbers and the growth of middle-class households are much greater than elsewhere in the United States, and there are nearly three-quarters of a million foreign-born households that are middle class (Table 3.8). The numbers are no less impressive in the other large immigrant regions: more than a third of a million in New York/New Jersey, and approaching that proportion in Texas/Arizona/New Mexico and in Florida.

There is no majority middle-class ethnic group in California, though Mexican-origin middle-class immigrants make up the single largest concentration—just over a third of all foreign-born middle-class households in California. But there are notable numbers of Filipino middle-class households, 75,377 (10.8% of the total), and a large number of

TABLE 3.8. The Distribution of Foreign-Born Middle-Class Households in 2000

	CA		NY/NJ		TX/AZ/NM		FL	
Hispanic	291,596	(41.7)	65,744	(19.5)	174,753	(61.5)	163,808	(56.2)
Asian	238,021	(34.1)	53,399	(15.9)	32,797	(11.5)	73,482	(25.2)
White	160,705	(23.0)	164,070	(48.7)	68,560	(24.1)	42,331	(14.5)
Other	8,681	(1.2)	53,447	(15.9)	8,099	(2.9)	12,013	(4.1)
Total	699,003		336,660		284,149		291,634	

Source: U.S. Bureau of the Census, Current Population Survey, 2000.

Russian/Eastern European immigrants who are middle class (11.6% of the total). It is a diverse and mixed middle class, a suggestion that middle-class status can extend beyond traditional backgrounds to encompass a very wide selection of households from different backgrounds. Perhaps it is the case that California encourages and supports greater opportunities than other locales. The other regions are also quite diverse in the makeup of their middle-class population, although Hispanics dominate in Texas/Arizona/New Mexico and are a majority in Florida.

The numbers are large and the growth has been substantial in three of the four regions (Table 3.9). While California is the largest in sheer numbers of foreign-born middle-class households, the growth rate of these households is striking in Texas and Florida too. No other state approaches California in the increase in raw numbers: more than a third of a million households in a 20-year period. It is clear that by dwelling on the influx of poor immigrants, the analysis is missing the transformation of the state with a rising professional/entrepreneurial middle class. Now foreign-born households are 24.3% of all middle-class households in California, and foreign-born households and the households headed by their children make up more than a third of all middle-class households in the state.

The story is quite different in New York/New Jersey—in numbers, in changes, and in composition. There are half as many middle-class households in New York/New Jersey as in California (Table 3.9). The largest part of the middle class has Russian/Eastern European origins, and more than a third of the foreign-born middle-class households are from these regions. At the same time there are substantial numbers of South American middle-class households (more than 27,000) and nearly 17,000 East Asian and Indian households that are middle class. New York/New Jersey stands out for the slow growth of foreign-born middle-class households and the nearly absolute losses in native-born middle-class households (Figure 3.8).

It is not simply that higher costs of living in New York compared to California explain the lower entry rates into the middle class, although for New York State, in particular, the special nature of the housing market does have an impact on the likelihood of the foreign born entering the middle class. The foreign born in New York State are concentrated in New York City, and Census data show that the rates of homeownership are lower in general in New York City than elsewhere in the nation. For

TABLE 3.9. Foreign-Born Middle-Class Households and their Proportion of the Total Middle Class by State

	1980		1990		2000	
California						
Asian	69,620	(2.8)	144,602	(5.5)	238,021	(8.2)
Hispanic	111,420	(4.5)	184,707	(7.1)	291,596	(10.1)
White	140,000	(5.7)	150,114	(5.7)	160,705	(5.6)
Other	3,660	(0.1)	5,660	(0.2)	8,681	(0.3)
Total	344,700		485,083		699,003	
New York/New Jersey						
Asian	20,220	(0.8)	46,315	(1.9)	53,399	(2.3)
Hispanic	29,000	(1.2)	62,972	(2.6)	65,744	(2.8)
White	211,900	(8.5)	179,911	(7.5)	164,070	(7.1)
Other	27,580	(1.1)	35,568	(1.5)	3,447	(2.3)
Total	288,700		324,766		336,660	
Texas/Arizona/New Mexico						
Asian	9,760	(0.5)	22,669	(1.0)	32,797	(1.2)
Hispanic	48,580	(2.4)	77,795	(3.5)	174,753	(6.5)
White	28,100	(1.4)	43,361	(1.9)	68,560	(2.5)
Other	1,478	(—)	3,298	(0.1)	8,099	(0.3)
Total	87,918		147,123		284,149	
Florida						
Asian	3,320	(1.2)	10,227	(0.7)	12,013	(0.7)
Hispanic	51,760	(4.4)	101,803	(6.6)	163,808	(9.7)
White	55,100	(4.7)	72,148	(4.7)	73,482	(4.3)
Other	4,480	(0.4)	18,011	(1.2)	42,331	(2.5)
Total	114,660		202,189		291,634	

Source: U.S. Bureau of the Census, Public Use Microdata Sample, 1980 and 1990, and Current Population Survey, 2000.

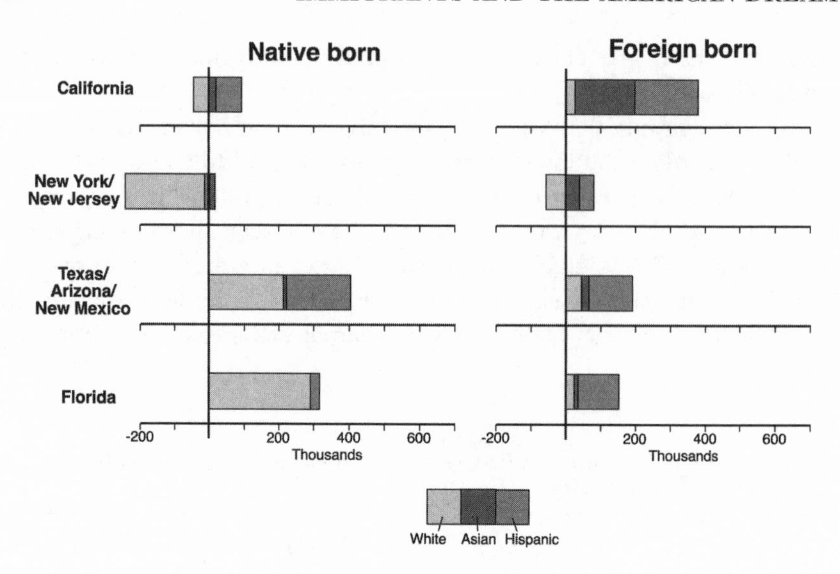

FIGURE 3.8. Growth in the middle class by state, 1980–2000. *Source:* U.S. Bureau of the Census, Public Use Microdata Sample, 1980, and Current Population Survey, 2000.

example, the homeownership rates for the population as a whole in 2000 are between 20 and 27% in Bronx County, Kings County (Brooklyn), and New York County (Manhattan), and are only 43% in Queens County. Only Richmond County (Staten Island) is near the national average. The effect of the lower general rates of ownership will have the effect of lowering the likelihood of being middle class in these central counties of New York State (i.e., the five counties—or boroughs—of New York City). Examining the likelihood of being middle class without homeownership provides a sense of the impact of the greater number who rent their homes in the central New York counties.

Overall, in the United States, the proportion of the foreign born with middle-class incomes is 36.2%. The proportion of middle-class households, income, and homeownership is 20.3%, a difference of about 16%. The difference between income alone and income plus homeownership is approximately 19% in New York State, which suggests that there is something on the order of a 3% decline that is attributable to using the homeownership measure in this special housing market. Or, expressing this in another way, the likelihood of a foreign-born household being middle class in New York is slightly underestimated if we use the combined income and homeownership measures.

Florida and Texas/Arizona/New Mexico have substantial similarities. They are both dominated by the Hispanic presence though with different compositions. Mexican foreign-born households made up more than half (51.7%) of all foreign-born middle-class households in Texas. At the same time, Southeast Asian foreign-born middle-class households numbered a substantial 21,254 and were nearly 8% of all foreign-born middle-class households. Despite the fact that Florida has a strong presence of foreign-born middle-class households, that presence is not totally Cuban. The latter make up less than 20% of all foreign-born households in Florida. Households with origins in South America (16.2%), Russia (12.5%), and Canada (5.7%) are important components of the middle class in Florida.

The patterns of growth for middle-class households are also different in Texas and Florida. Both regions had strong foreign-born growth *and* strong native-born growth. Here we see the combination of economic opportunities (Texas) and educational background (Cuban immigrants in Florida) influencing their transformation into the middle class (Figure 3.8).

The geographic variability is important. It is important to move beyond only national patterns, as those patterns vary remarkably from state to state. To reiterate, California's middle-class growth was almost entirely foreign born, whereas in Texas and Florida the native-born ethnic population still contributes to most of the middle-class growth. New York/New Jersey has only modest middle-class growth, and even that growth is largely from foreign-born increases. While as a whole foreign-born households contributed 28% of the growth in middle-class households between 1980 and 2000 in the United States, they contributed 90% of that growth in California, 100% in New York/New Jersey (though a small number, of course), and approximately a third of the growth in Texas and Florida (Table 3.10). Clearly, the foreign-born population is changing the composition of the U.S. middle class.

It is notable that the growth in the middle class goes well beyond the traditional major ethnic inflows. White and other foreign-born middle-class households have also increased their presence in the past two decades. New York/New Jersey and California have the largest numbers, and they are the biggest proportion of all middle-class households in any of the four regions being examined. In fact, foreign-born plus native-born Hispanic and Asian households who are middle class constitute a fraction under 35% of all middle-class households in California. Clearly, very large numbers of immigrants are thriving in their new homes. New York

TABLE 3.10. Proportion of Middle-Class Growth That Is Foreign Born

United States	28.7
California	90.1
New York/New Jersey	99.5
Texas/Arizona/New Mexico	29.7
Florida	33.7

Source: U.S. Bureau of the Census, Public Use Microdata Sample, 1980, and Current Population Survey, 2000.

has a significant middle-class presence of foreign-born white middle-class households, and there are quite large numbers of "other" foreign-born households in New York and Florida. Even though Hispanics by and large dominate the new foreign-born middle class, there are large numbers of other ethnic groups in the regional distributions.

The percentage increase in middle-class numbers is quite variable, both by time and place (Table 3.11). There were losses in New York/New Jersey, but very large percentage increases in Florida and Texas/Arizona/New Mexico. The proportionate gains are greater for the foreign-born population than the native-born population, but this is largely a function of the quite small initial populations. Over the two-decade period, the

TABLE 3.11. Percentage Change in Middle-Class Households by State

	All		White		Hispanic		Asian	
	NB	FB	NB	FB	NB	FB	NB	FB
California	1.9	115.3	–2.6	14.8	35.4	161.7	31.8	241.9
New York/ New Jersey	–9.6	16.6	–12.0	–22.6	–6.8	126.7	220.2	164.1
Texas/ Arizona/ New Mexico	23.8	223.3	13.4	144.0	79.8	259.7	55.5	236.0
Florida	33.1	153.6	23.3	33.4	99.6	216.5	—[a]	261.8

Note. NB, native born; FB, foreign born.
Source: U.S. Bureau of the Census, Public Use Microdata Sample, 1980, and Current Population Survey, 2000.
[a]Small sample size.

middle-income and middle-class groups have nearly all doubled (with exceptions in New York/New Jersey); in Florida, in some instances, they have tripled. In Texas, the increase in the total foreign-born middle class exceeded 200%. The gains of the Hispanic foreign-born middle class in two decades in Texas were 260%. The proportion of *growth* in the middle class that was taken up by the foreign born varied from 28.7% for the United States as a whole to nearly the whole amount in New York/New Jersey and California.

Proportions

The patterns across states are complex, although there are some general threads that can be drawn out of the diagram of proportions of change (Figure 3.9). The proportion of foreign-born Asian households in the middle class declines except in New York/New Jersey, but it is at very low levels there in any event. Hispanic foreign-born households are lower in their levels of gains than are Asian or white foreign-born groups, but in Texas they actually increased their gains in the 1990s. White foreign-born households have relatively high rates of participation in the middle class and are more nearly stable in their levels over time.

The stories for the native-born ethnic groups by states are really extremely positive in three of the four regions, especially for Hispanics. In Florida, the native-born Hispanic households are holding their own over time and have rates of middle-class gains that are similar to rates for the society as a whole. In Florida, the relative proportion of the Hispanic population which is middle class has not declined. Even more notable is the fact that the foreign-born population in Florida has middle-class entry rates which are like the native-born entry rates in California and Texas, and two and a half times higher than those of the native-born Hispanics in New York/New Jersey. This finding emphasizes the regional variation in middle-class gains and highlights the findings for California and Texas, where indeed there is strong evidence of an emerging middle-class Hispanic population.[15]

At least a part of the explanation for the strong differences in Hispanic middle class entry is related to the varying composition of the Hispanic population in these states. The New York/New Jersey area has very large proportions of Dominican and Puerto Rican populations, and Florida is dominated by Cuban-origin populations. The contrast between Florida and New York/New Jersey is clearly a contrast between relatively well-educated Cuban-origin populations in the former and much less

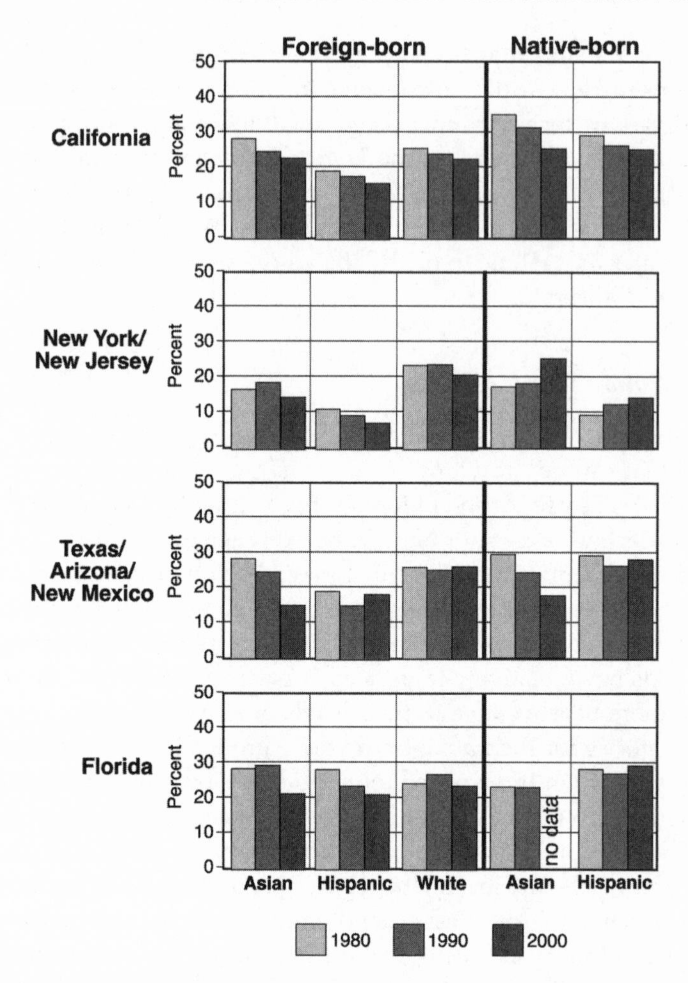

FIGURE 3.9. Middle-class households as a proportion of all households by region. *Source:* U.S. Bureau of the Census, Public Use Microdata Sample, 1980 and 1990, and Current Population Survey, 2000.

well-educated and lower-skilled populations from the Dominican Republic and from Puerto Rico in the latter. About 15% of the New York/New Jersey Hispanic population is from the Dominican Republic, and another 30% is from Puerto Rico. In contrast, Florida's Hispanic population is approximately 55% Cuban. The findings about middle-class penetration emphasize again the important role of human capital, of the education and skills which immigrants bring with them or which they acquire after they arrive in the United States. The findings also raise the central question of the future trajectories of those without the human capital or with large families to support.

The other part of understanding the rates of middle-class gains is connected to the interrelated issues of large numbers of recent immigrants and the cost of homeownership. Clearly, in expensive housing markets such as San Diego, San Francisco, and Los Angeles, the ability to become a homeowner is curtailed by cost, as reflected in the decreasing proportion of households who are middle class in 2000. In addition, as I pointed out in an earlier part of this chapter, the large number of very recent immigrants has increased dramatically the base on which we are calculating the number of households who are middle class.

TIMING AND ECONOMIC CONTEXTS

The pace of middle-class entry is also influenced by the year of arrival in the United States (Figure 3.10). As expected, earlier arrivals are more likely to be members of the middle class than are later arrivals. If nothing else, it takes time to acquire occupational skills or translate existing skills to good jobs, buy homes, and become established. There is a noticeable gap between immigrants who arrived before 1980 and those arriving later. The step-down in the proportions of households who are middle class occurs fairly regularly after 1980, but again there are regional variations. In California and New York/New Jersey, the step-down for Asian households does not occur until about 1990. For Hispanic households in Florida, the step-down is only after 1990. Again the results reiterate the way in which differing immigrant compositions (e.g., the recent arrivals after 1980 in California) are affecting the base of the population. There are fewer households that can make the upward mobility shift. At the same time the sanguine view will be that they too, perhaps at a lower rate, will achieve middle-class status.

The most recent arrivals are the least likely to be in the middle class. It is this finding which explains the overall lower level of middle-class entry by the foreign born. In turn, this leads to the question of whether 20 years is a meaningful threshold, and whether it will require at least this long, or longer, for the foreign born to make the transition to the middle class. Will the post-1980 arrivals have a much more difficult time of making it to the middle class at all? If there are marked differences in skills between those who arrived before 1980 who have joined the middle class and those who arrived later, then the trajectories of the two groups may

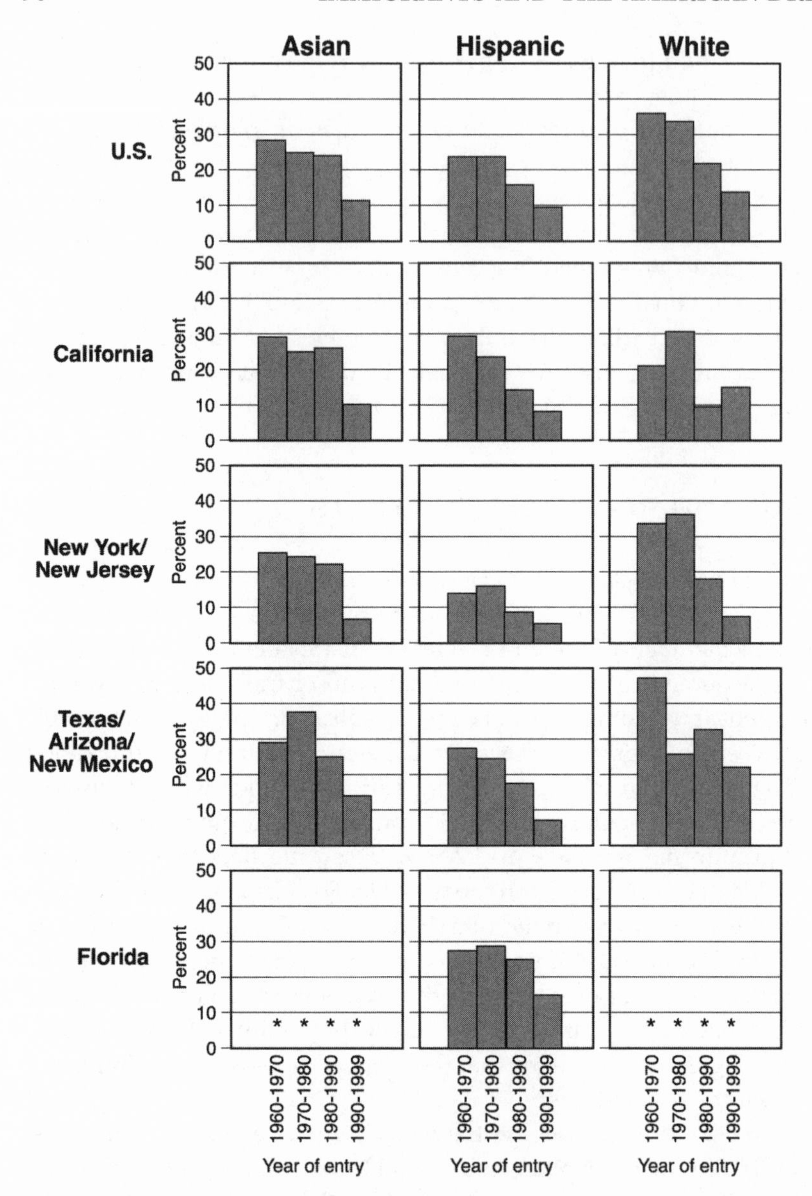

FIGURE 3.10. The proportion of foreign-born households who are middle class by year and entry into the United States. *Source:* U.S. Bureau of the Census, Current Population Survey, 2000.

not be parallel and may continue to diverge over time. It is not possible to provide a complete answer to the question until the full data from the 2000 U.S. Census are released, and it is just this question which is at the heart of the continuing debate about immigrant progress. Some of the cohort data discussed in this book provide a possible set of scenarios of what may happen.

There are substantial differences across the states and some interesting reversals in the rates of middle-class entry by time of arrival. Florida stands out for the generally very high levels of entry to the middle class and especially for the older cohorts, and the drop-off in levels of entry occurs only after 1990. Again, the levels of entry are visibly lower in New York/New Jersey. In the context of the average levels of middle-class participation in the United States as a whole, the outcomes for three of the four regions are quite notable. All regions except New York/New Jersey outperform the overall U.S. average for arrivals before 1980. Those who arrived in the United States earlier have made important strides in joining the middle class, and their proportions are not distinguishable from the total population. Florida is notable, especially with the proportions of the older age cohorts. These findings emphasize that socioeconomic integration appears to be working, and an optimistic view is to expect that the later arrivals may well also move up and join the middle class.

SUMMARY AND OBSERVATIONS

The evidence given in this chapter sets out a compelling case that immigrants have made progress. The results are consistent with research data examined from an earnings perspective that also suggests significant progress (Duleep and Dowhan, 2002).

In the context of the middle class as a whole, foreign-born households are doing well. If we add the ethnic native-born groups and the foreign born together (Hispanic and Asian households, respectively), they are now a very large proportion of all middle-class households, more than 34% of the middle-class population in California and nearly 28% in Texas. The proportion is nearly 20% in Florida, and even in New York/New Jersey it is more than 16%. The proportions are high for the foreign-born population in California and Florida, though as the Hispanic population is a very large proportion of all foreign born in Florida and California, these proportions are not equivalent. True, these groups have not reached parity with their proportion of the total population, but the numbers are moving toward parity. Time seems to be on the side of the immigrant middle class, and it seems likely, as the following chapters show, that they will weather the growing inequality in U.S. society and continue their socioeconomic gains in the coming decades.

NOTES

1. Self-improvement/self-help book titles ranging from estate planning and money management to personal life strategies fill several shelves in most local bookstores, and new titles are being added constantly.
2. The middle-income range is from $34,058 to $84,975 in 2000 dollars. Household incomes for earlier periods are adjusted by the Consumer Price Index (CPI).
3. Housing comprises about half of the net worth of the average homeowner (G. McCarthy, Van Zandt, and Rohe, 2001). Housing is a more significant share of total household net worth for minorities: about 61% for Hispanics.
4. To reiterate an earlier note, using a household measure of income rather than a family measure of income is a slightly more generous definition of the income necessary to be middle class, as more individuals in that household might contribute to the total income available.
5. On one hand, the notion of using alternative measures of middle-class status—median income or homeownership—resonates with our notions of the way in which households make it up the socieoeconomic ladder. Under this definition, households may enjoy some of the facets of middle-class status by owning a home even if they do not have an income that is in the middle-income range. On the other hand, using an "either/or measure" may be capturing households who are in the very low end of the homeownership market, housing which may be in poorer neighborhoods and inner-city areas, without the concomitant associates of the middle-class lifestyle.
6. Because we are using the Current Population Survey (CPS) for 2000, strictly speaking the age range is 40–48, as the income data is collected for the previous year, in this case 1999. But because we are using proportions, the results will be influenced marginally if at all.
7. "The 1.5 generation" is the term given to those immigrants who arrive as children, usually under 5 years of age, though in this case the group would be from 10–19 years old in 1980 if they are 20–29 in 1990.
8. There is always the question of what is the most appropriate comparison group against which to compare the progress of the immigrant cohorts. I have chosen to use all native born as the overall comparison and the relevant

native-born group for each of the comparisons for Hispanic, Asian, and white foreign-born groups. However, in the end it is the actual, nearly always positive trajectories of progress for the immigrant groups that are equally important to the extent to which they are catching up to the native born.

9. The notion of the model minority, perhaps a stereotype, is that of a hardworking group "fitting in" and using the system, especially education, to get ahead in American society. For some, even the term "Asian American" may blur what is a very diverse group (Cheng and Yang, 1996; Ong and Hee, 1994).
10. Occasionally throughout the text I use "New York," or "Texas" as a shorthand for New York/New Jersey or Texas/Arizona/New Mexico, respectively, but in every case the analysis is for the combined region.
11. To the extent that the definition of the term "middle class" is dependent on homeownership and that in general rates of homeownership are somewhat lower in New York (especially New York City), the results can be discounted somewhat. However, the fact that native-born whites in New York/New Jersey have much higher rates of middle-class entry suggests that it is not simply a homeownership issue.
12. In some cases the rates were considerably above 50%, a remarkable level of participation in its own right, but they are often based on small numbers, and I truncated the graphs at 50%.
13. For a comparative perspective on the second generation in Canada, see Boyd (1998).
14. The models are simple logit functions of the probability of being in the middle class (income and homeownership) as a function of the independent variables. The data are taken from the 1990 Public Use Microdata. The data for 2000 have not yet been released, but it is unlikely that the general findings will change.
15. The results are not markedly affected by the use of a national income range. The median is slightly higher in California, but overall the range is sufficiently large that it does not introduce a bias.

CHAPTER 4

★ ★ ★

Entering the Professions

Consistent with the theme of whether and how immigrants are making it to middle-class status, this chapter asks the following questions: (1) Are immigrants becoming members of the most desirable occupations and industries? (2) Which immigrants are the most successful at high-status occupational gains? These are different questions than those which have motivated much prior research about wage gaps between the native-born and foreign-born populations.[1]

To elucidate these complex issues, I examine occupational gains over the past two decades and how they vary by immigrant origin and across regions in the United States. The analysis is mainly cross sectional, comparing levels of occupational success at different points in time. It concentrates on the "successful" immigrants rather than the large pool of low-skilled immigrants who are working in low-wage jobs. Briefly, some groups of immigrants are making significant inroads into high-status occupations. At the same time, the same bifurcation which seems to be occurring in U.S. society at large may also be shaping the trajectories of immigrant occupational gains.

The same debate about income gains that was reviewed in earlier chapters—as to whether immigrants will catch up or whether they will fall further behind—is also a part of the discussion of occupational achievements. But in this chapter the focus is not so much on the relative progress of immigrants as it is on what we can say of the immigrants who are successful. Who are they, how have they succeeded in the U.S. labor market, and what insights can we gain from their successes? The analysis here is focused on changes in occupational status and on whether or not the foreign born are moving up the occupational ladder. Certainly, it is important to be able to say that "I am a manager" rather than "I wash dishes" even if the managing is in the local fast-food restaurant.

THE CONTEXT OF UPWARD OCCUPATIONAL MOBILITY

In all societies, occupations are arranged hierarchically. For that reason, one's occupation and one's social position reflect one another (Miller, 1991). In the United States, the highest status is accorded perhaps to Supreme Court Justices and leading figures in universities and scientific research groups; among the lowest-status occupations are busboys, kitchen helpers, and a myriad of low-skilled service activities. More generally, high-skilled and information-based occupations are at the apex, and service and manual-laboring activities at the bottom. In this structure workers with fewer skills fill low-level occupations and work in industries that have large numbers of these jobs.

Although hierarchies of occupations are relatively fixed, people can enjoy social mobility upward, either personally or vicariously through their children. Children of parents in lower-skilled occupations move up as they acquire the educational skills to occupy higher levels in the occupational hierarchy (Treiman and Ganzeboom, 1990; Ganzeboom, Treiman, and Uptee, 1991). Of course, not all immigrants surpass their previous generations, and increasingly immigrants are arriving with professional skills, making it a more complex picture than in the past. Even so, social mobility is as much at the heart of the immigrant American Dream (and associated economic gains) as the material gains themselves.

This chapter looks beyond the narrow issue of comparative wage levels to the issue of immigrant gains in particular occupations and industries. Are immigrants successfully penetrating professional and specialized occupations and industries? By focusing on the successes in the entry to professional and administrative occupations in particular, the analysis counters the focus on wage trajectories alone. Whatever wage gap exists, the successful entry to professional and technical jobs is a central part of the upward mobility of immigrants, and in turn for their children.

Previous Findings

Most of the previous work on what I am calling industry and occupation penetration has been in the context of niches (Model, 1985; Rosenfeld and Tienda, 1999; Waldinger and Bozorgmehr, 1996). A niche occurs when an ethnic group is more strongly represented in a particular occupation than the average of all ethnic groups. Thus, if 80% of Vietnamese workers are in the electronics industry, we would say that it is a niche occupation. There are countless anecdotal examples: Korean dry-cleaning businesses, budget motels run by Asian Indians, doughnut shops run by Cambodian immigrants, and nannies from El Salvador.[2] The emergence of niches is explained as an outcome of ethnic group networks of mutual support. When immigrants dominate a particular industrial sector and when they can use their network ability for mutual support, they often achieve success levels greater than would occur otherwise (Waldinger, 2001). Of course, niches also arise because entry into some industries is denied to new arrivals, and niches change over time. In California, agriculture was once dominated by Japanese immigrants, but now the new Mexican immigrants make up almost 80% of the employment in agriculture in that state (Allen and Turner, 1997; Clark, 1998). Immigrant niches vary across metropolitan areas too and are, for example, much more varied and significant in Los Angeles than in other large metropolitan areas (Logan, Alba, and McNulty, 1994).

The studies of immigrant niches have not focused as much on the *process of penetration or gains,* of the change in the numbers in management, or in the professions, over time. Previous studies of immigrant concentrations merely note the greater concentration of immigrants broadly in service-sector occupations and in some manufacturing sectors. However, a 1997 National Academy of Sciences (NAS) study did draw attention to the increasing numbers of immigrants with high levels of education that are in high-status occupations (Smith and Edmonston, 1997, pp. 209–219). At the same time they reiterated the oft-cited finding that immigrants are often laborers, fabricators, or work in the service sector, because they possess fewer of the skills required for success in the U.S. labor market. As a result they tend to have lower incomes than their native-born counterparts (Garvey, 1997). Yet some previous research has already shown that, at least in California, immigrants have made major inroads into almost all occupational/industrial sectors but the public administration sector (Clark, 1998).

Case studies in Los Angeles have established also that there has been significant entry into the retail sector by both foreign-born Hispanics and Asians (Wright and Ellis, 1997). Although these studies focus on immigrant channeling, they do provide important observations about the tendency of groups to achieve a critical mass in certain industries. These studies also emphasize the role of human capital in creating immigrant concentrations in particular industry groupings. Using a cohort analysis, Myers and Crawford (1998) show that after controls for age and year of arrival, human capital is a critical factor in how immigrants come to be allocated to lower- or higher-level occupations. Legal status is also an important factor in the likelihood of making gains in higher level occupations (Wright, Bailey, Miyares, and Mountz, 2000).

Immigrant entrepreneurship stands out as one path to success (Light and Bhachu, 1993). Self-employment appears to be an important mechanism that allows workers in particular sectors to earn a considerable bonus above what they might make in regular employment with the same skills (Waldinger and Bozorgmehr, 1996). Koreans, for example, have been among those most able to use the self-employment approach to successful upward trajectories in the labor market. At the same time, as Abelmann and Lee (1995) point out, Koreans often used the entrepreneurial route because U.S. employers did not credit their degrees, even from prestigious South Korean universities. Armenians and Russians also have high rates of self-employment (Light and Roach, 1996). In contrast, industries and occupations that are penetrated by Central Americans and Mexicans are often occupational traps with little prospect for upward mobility. There is a well-developed literature on immigrant niches, the industries and occupations that are dominated by different immigrant groups; there is much less research on general industry and occupation entry rates. Two important issues are less well studied: (1) Are immigrants making gains in the upper levels of occupations and industries, and—if so—who are they? (2) How many immigrants are trapped in low-end service activities, and who are they? Yet, in the argument in this book these questions are central to the development of a new immigrant middle class.

Recall that our focus in this research is not on channeling (the way in which immigrants are "directed" to particular occupations) or on niches per se. Rather, the interest is on the dual issues of whether some groups are being locked into some industries and occupations, and other groups are able to access a wider range of industries and occupations and how those industries and occupations are changing over time. Clearly the focus is a subtle variation of immigrant niches but an important one, as it emphasizes access and upward mobility rather than channeling and niche behavior.

THE CHANGING U.S. LABOR FORCE

The large-scale immigration to the United States in the 1980s and 1990s coincided with a generally expanding U.S. economy, especially during the latter half of the 1990s. The U.S. economy added more than 30 million new jobs and, by extension, job opportunities in the two-decade period from 1980 to 2000. Even though the economy slowed at the beginning of the 21st century, this long-term growth has provided a large number of new jobs for both the native born and new immigrants alike.

Both the native-born and the immigrant workforce expanded, but while the native-born workforce grew by about 27% and some 24 million workers, the foreign-born workforce tripled from a little more than 6 million to nearly 18 million in the same period (Figure 4.1). In 1980 immigrants made up about 7% of the workforce. The proportion increased to nearly 12% in 1990 and 16% in 2000. Immigrants accounted for one-third of the jobs added to the economy.[3]

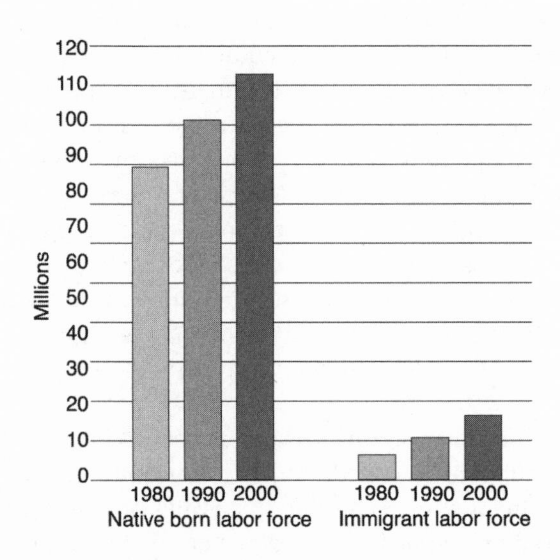

FIGURE 4.1. Change in the U.S. labor force, 1980–2000. *Source:* U.S. Bureau of the Census, Public Use Microdata Sample, 1980 and 1990, and Current Population Survey, 2000.

Not only did the foreign-born presence in the U.S. workforce increase in the last two decades of the 20th century, both foreign-born and native-born women became a much larger presence in the labor force (Table 4.1). The increases were general across all foreign-origin groups—hardly surprising given the continuing high-level immigration flows and the labor-market opportunities in the United States. However, there are two remarkable and important findings which are illustrated in a graph of the immigrant workforce by ethnic origin: the quadrupling of the Mexican-born labor force, from a million to more than 4 million, and the great diversity in foreign-born contributions to the U.S. labor force (Figure 4.2). Even though the foreign-born workforce is dominated by the combination of Mexican and Central American immigrants, about 32% of the total workforce, they are still less than a third of all foreign born in the U.S. labor force. The "other" category is actually larger than the number of Mexican workers. Equally interesting is the finding that the "other" category—immigrants from Africa, Australia, New Zealand, and the Pacific Island nations—are growing as rapidly as any individual group.

The data show broad immigrant gains in every industrial sector (Figure 4.3). Although they dominate some sectors in proportional terms, there are very large numbers of immigrants in almost all industrial sectors. While the numbers are large and growing in wholesale/retail and in manufacturing, the numbers are large in the professions too. Clearly, they are not, contrary to perceptions, confined to employment in agriculture, service activities, or manufacturing. It is true that there were very large percentage increases in the construction sector, but the percentage increases in finance and business were almost as large, and numerically they were twice as big (Table 4.2). The foreign-born proportions of the

TABLE 4.1. Employment in the U.S. Labor Force

	Foreign born			Native born		
	Men	Women	Total	Men	Women	Total
1980	3,576,400	2,624,300	6,200,700	50,280,700	37,819,420	88,100,120
1990	6,771,163	4,906,421	11,677,584	53,708,815	46,492,283	100,201,098
2000	10,515,991	7,367,568	17,883,559	58,587,747	53,968,214	112,555,961
% change 1980–2000	194.0	180.7	188.4	16.5	42.7	27.8

Source: U.S. Bureau of the Census, Public Use Microdata Sample, 1980 and 1990, and Current Population Survey, 2000.

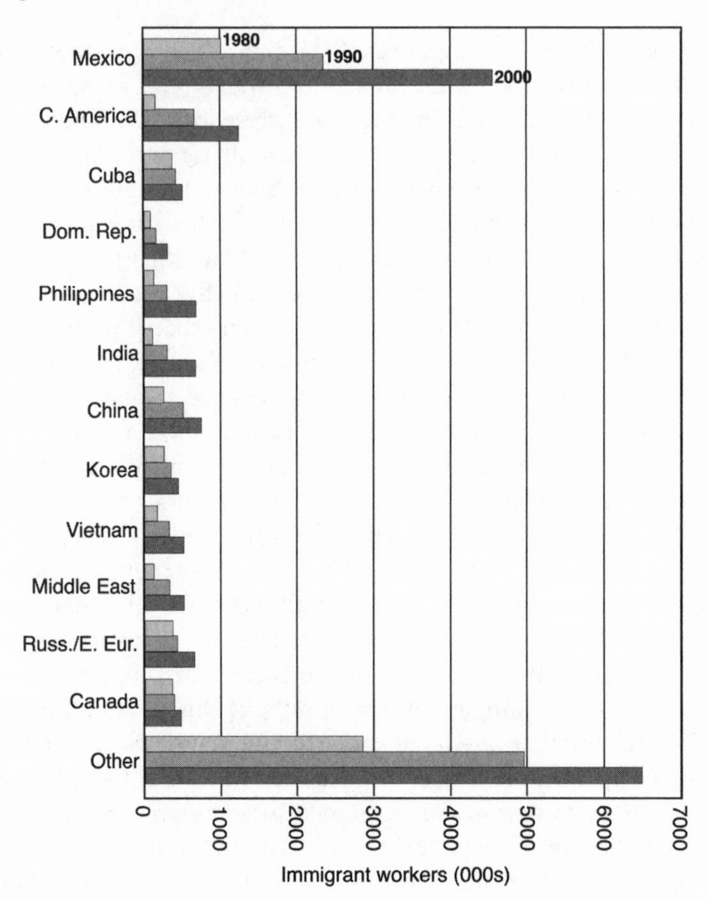

FIGURE 4.2. U.S. labor force growth of the foreign born by ethnic origin. *Source:* U.S. Bureau of the Census, Public Use Microdata Sample, 1980 and 1990, and Current Population Survey, 2000.

professional, finance, and public administration occupations are remarkable given how recently they have arrived.

The foreign-born workforce is now a major part of the economy in all regions in the United States. Consistent with our snapshot of immigrants in 2000, the immigrant workforce is largest in California and has been growing rapidly (Figure 4.4). The native-born labor force continues to expand in California, in Texas and nearby Southwestern states, and in Florida, but is stable in New York/New Jersey. The native-born growth is especially vigorous in Texas and to a lesser extent in Florida and Califor-

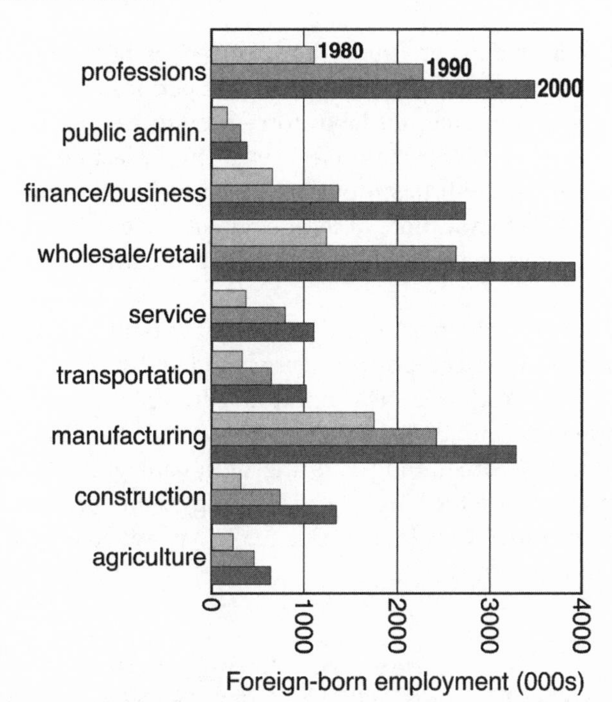

FIGURE 4.3. Change in foreign-born employment by industry in the United States, 1980, 1990, and 2000. *Source:* U.S. Bureau of the Census, Public Use Microdata Sample, 1980 and 1990, and Current Population Survey, 2000.

TABLE 4.2. Change in the Native-Born and Immigrant Labor Force in 1980–2000 by Industry in the United States

	NB	FB	% change NB	% change FB
Professions	10,832,214	2,334,343	60.2	207.0
Public Administration	737,097	218,466	15.3	123.3
Finance–Business	6,859,666	2,070,129	76.8	315.6
Wholesale/retail	4,818,905	2,663,334	26.9	212.5
Service	1,207,889	738,659	35.5	200.2
Transportation	1,785,025	698,981	27.0	217.3
Manufacturing	–2,933,116	1,520,677	–14.9	86.5
Construction	1,938,271	1,024,596	36.9	327.9
Agriculture	–790,110	413,674	–22.9	184.1

Note. NB, native born; FB, foreign born.
Source: U.S. Bureau of the Census, Public Use Microdata Sample, 1980, and Current Population Survey, 2000.

nia, but it is clear that the immigrant presence is now important in all these states. In every state, for the two-decade periods examined here, the contribution of the immigrant labor force growth has been at least 30%, and in California and New York/New Jersey in the last decade the immigrant labor force growth has provided much of the growth in the labor force (Figure 4.5). In addition, at least some of the growth in the native-born labor force is made up of the children of earlier waves of immigrants.

The foreign-born workforce is a complex mix of ethnic origins and backgrounds, and the composition varies widely by state (Table 4.3). In Florida and New York, the most important finding may be that no one group dominates. Only in California and Texas are Mexican and Central American foreign-born immigrants the dominant group in each state. Asian immigrants are the single largest group of the foreign-born workforce in New York/New Jersey and make up about a quarter of the workforce in California.

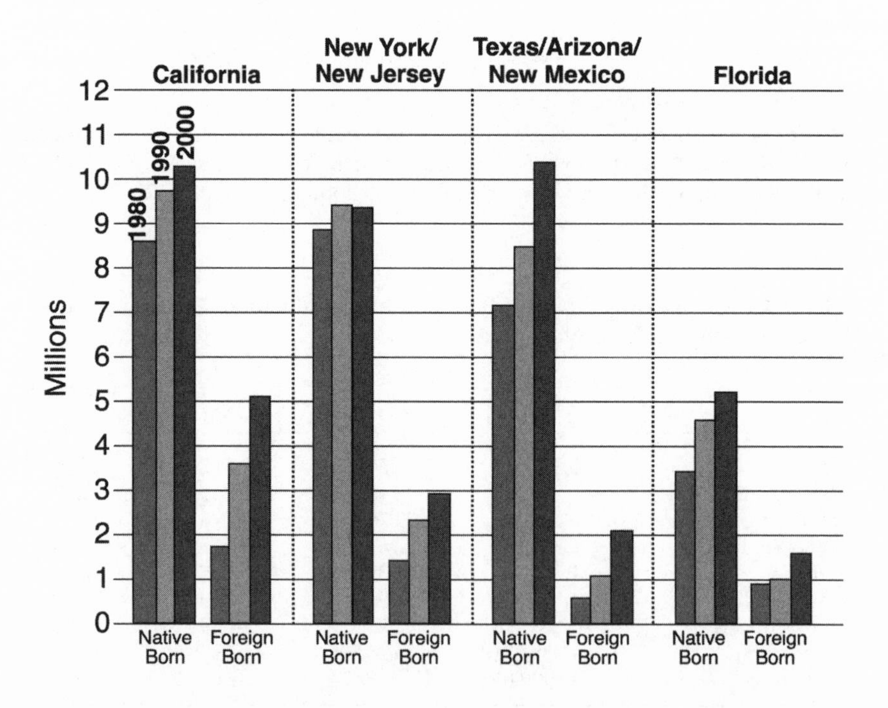

FIGURE 4.4. The foreign-born labor force in large-immigrant-impact states, 1980–2000. *Source:* U.S. Bureau of the Census, Public Use Microdata Sample, 1980 and 1990, and Current Population Survey, 2000.

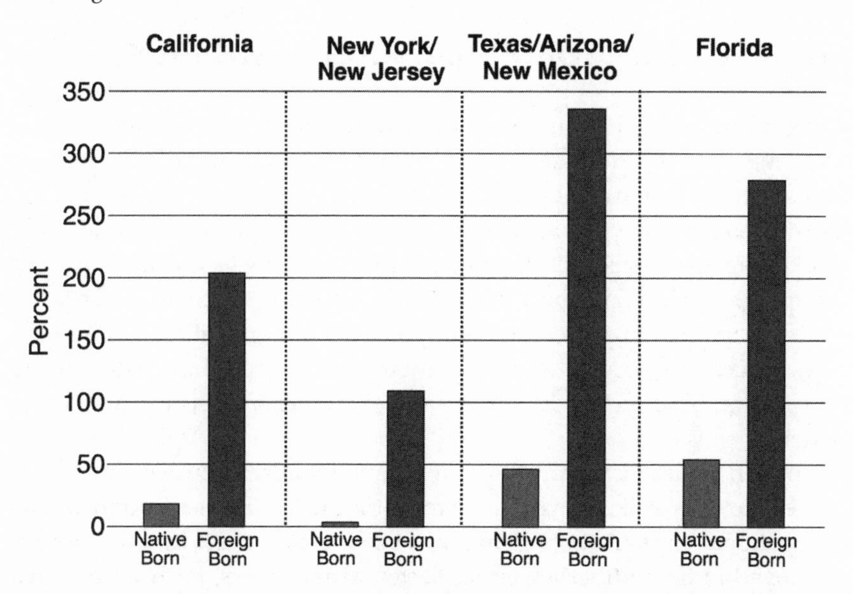

FIGURE 4.5. Percentage increases in the native-born and foreign-born labor force for large-immigrant-impact states, 1980–2000. *Source:* U.S. Bureau of the Census, Public Use Microdata Sample, 1980 and 1990, and Current Population Survey, 2000.

TABLE 4.3. Composition of the Immigrant Workforce by Origin in 2000

	Mexico/ Central America	Cuba	Asia	Dominican Republic	Russia/ Eastern Europe	Middle East
California	49.5	—[a]	24.1	—[a]	2.2	4.2
New York/New Jersey	9.2	0.8	14.3	7.0	7.2	3.2
Texas/Arizona/New Mexico	66.0	—[a]	9.8	—[a]	0.6	1.8
Florida	16.3	19.3	4.7	1.4	1.5	1.5

Source: U.S. Bureau of the Census and Current Population Survey, 2000.
[a]Small sample sizes.

THE EVIDENCE FOR PROFESSIONAL ADVANCE

Occupations form five broad socioeconomic groupings, defined by type and level of skill. Professional and managerial workers, one of the two white-collar occupational groups, include managers, administrators, scientists, teachers, doctors, and nurses. The second group of white-collar workers includes sales clerks, clerical workers, and technicians. A third grouping, of what have been traditionally called blue-collar workers, includes craftsmen, machine and equipment operators, laborers, and handlers. Fourth are service workers, ranging from cooks and custodians to barbers and beauticians. Finally, there are farmers and farm-related occupations.

When we are using this kind of occupational structure, we must recognize the difference between a sector (or industrial) classification and an occupational classification. While most white-collar service workers are employed in the industrial service sector, as managers, for example, that sector also includes many other blue-collar workers such as telephone repair workers and auto mechanics. Also, the service sector includes many professionals including doctors and lawyers. In the remainder of the discussion in this chapter the focus is on *occupations* rather than industries, as the concern is with what is happening to *people* rather than the economy per se.

The number of immigrants has increased significantly in all occupations (Figure 4.6), but the most powerful finding is that the number of

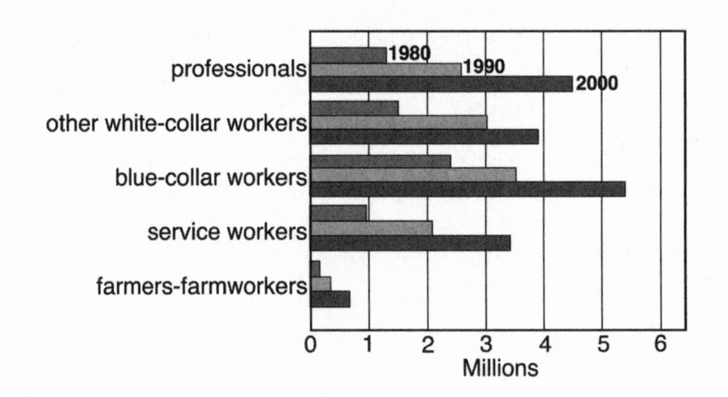

FIGURE 4.6. Immigrant employment levels by occupational categories in the U.S. labor market, 1980, 1990, and 2000. *Source:* U.S. Bureau of the Census, Public Use Microdata Sample, 1980 and 1990, and Current Population Survey, 2000.

professional and other white-collar workers in combination is larger than the number of blue-collar workers. In 2000 there were almost twice as many foreign-born professionals and other white-collar occupations as there were blue-collar workers (Figure 4.6). Not unexpectedly, because so many of the most recent immigrants are low skilled, there are large increases in the lower-wage manufacturing sectors, among operators and laborers in blue-collar occupations. Even so, the numerical increase in the professions is actually greater. We hear a great deal about immigrant farm workers, but they are only a small proportion of all workers, even if they are disproportionately dominated by immigrants.

To gauge the progress of immigrants into the higher-status jobs, this chapter examines the raw size of the immigrant professional workforce and the proportions that they make up of the major professions. In addition, an index of penetration, defined as the ratio of the percentage increase in the foreign born who are professionals, divided by the percentage increase in the total immigrant labor force, is used to examine their relative gains over time. In other words, were the gains in the professions greater or less than the overall increase in the immigrant workforce?

The absolute increase in the *numbers* in the professions was greatest within the managerial group and "other" professionals—engineers, scientists, and computer specialists, for example—but the increases in those in medical occupations and education were not insignificant (Figure 4.7). There is clear evidence of professional occupational gains. Taking managers as an example, their number more than tripled in the two-decade

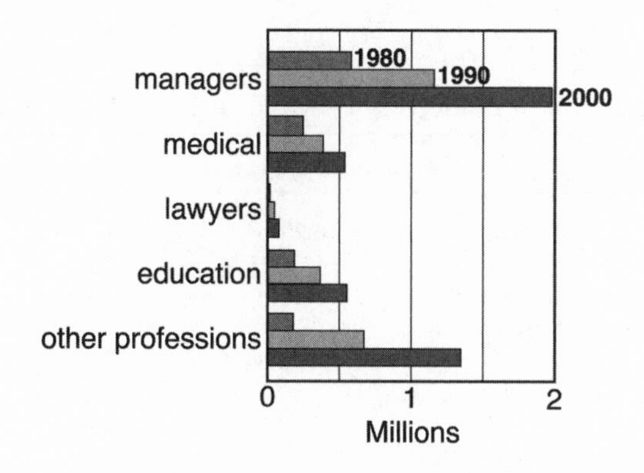

FIGURE 4.7. Size of the immigrant labor force by professional occupation. *Source:* U.S. Bureau of the Census, Public Use Microdata Sample, 1980 and 1990, and Current Population Survey, 2000.

period, and immigrants accounted for 15% of the increase in managerial occupations. The number and growth in the legal area is lower. Opportunities for new immigrants to create legal services for fellow immigrants will emerge, but because legal practice is nation specific there will be slow entry into the legal professions for those who have been trained elsewhere.

The *proportions* in all professional categories increased in the two-decade period (Table 4.4). Now more than 14% of jobs in the medical professions and in the engineering and science professions are held by the foreign born. In addition, it has been a relatively rapid increase, nearly doubling in percentage terms in 20 years. Recall, too, that the immigrant proportion in the labor force as a whole is about 13.7% in 2000. The proportions in medical specialties are notable. They include nurses and hospital workers as well as doctors, and the increase went from 9.7 to 14.4% for all medical professionals. On a visit to the hospital or a doctor's office, the average patient has a nearly 1 in 5 chance of being attended by a foreign-born nurse or doctor.

Foreign-born physicians numbered 119,108 in 1990 and made up 20% of all physicians in the United States (Bouvier and Jenks, 1998). The number has certainly grown since then. Foreign-born doctors are a larger proportion of all physicians in the United States than they are of the total U.S. population. Nearly half of the foreign-born physicians are Asian, and another 35% are non-Hispanic whites (Bouvier and Jenks, 1998). Many of the foreign-born doctors are women, and among Asian foreign born they are likely to be young women. The number of foreign-born nurses has

TABLE 4.4. Foreign Born in the U.S. Professional Labor Force, 1980–2000

	1980	1990	2000
Total foreign-born labor force	6,200,700	11,677,584	17,883,558
Total professional labor force	1,328,100	2,626,709	4,507,268
Percent of total labor force	5.9	8.6	11.7
Percent of occupational groups			
Managers	5.8	8.4	10.3
Medical professions	9.7	11.5	14.4
Lawyers	2.6	4.4	6.8
Educators	4.0	6.6	8.8
Other professions	7.6	10.6	14.6

Source: U.S. Bureau of the Census, Public Use Microdata Sample, 1980 and 1990, and Current Population Survey, 2000.

grown too. In 1990 the 167,000 foreign-born nurses were largely Asian. As in the case of doctors, 44% of the foreign-born nurses were from Asian countries, and nearly 90% of all Asian nurses were foreign born. However, there is a sizable proportion of Hispanic foreign-born nurses—nearly 11% of the foreign-born nurses were Latino. Even so, Asians are about four times more likely to be nurses as are Hispanics, a finding that is attributable to the overwhelming presence of Filipinos in the nursing and health professions generally. The physicians and nurses are likely to have good incomes and are the quintessential members of the new immigrant middle class. Large numbers have some higher education, and many have college degrees from their countries of origin or gained in the United States.

A simple index of penetration, the percentage gain in the professions as a ratio of the percentage gain in the labor force, provides additional support for our general argument of increasing upward movement in status (Table 4.5). The index for total professionals and for almost all the divisions is above 1.0. In addition, in several instances the indexes increased for the most recent decade. That is, immigrants are entering professions at a rate that is nearly always greater than their entry to the labor force in general. It is true that in some cases—law, for instance—the ratios are influenced by the relatively low initial numbers, and in education they are somewhat below the ratio of parity. However, the positive story, generated from examining absolute numbers and proportions, holds up in a comparative analysis of flows into the labor force.

Thus far the results are consistent with the general argument of upward occupational gains—a positive story of immigrant progress in a changing postindustrial labor market. The question that must now be examined is the extent to which the story of successful occupational gains is true across immigrant groups. Clearly, following the arguments of upward occupational mobility, we would expect that those groups with

TABLE 4.5. Professional Penetration of Occupations by the Foreign Born

	All professions	Managers	Medical	Legal	Education	Other
1980–1990	1.1	1.2	0.8	1.8	1.1	1.2
1990–2000	1.4	1.4	0.8	1.8	1.0	1.9

Note. The index is defined as the ratio of the percentage gain in professions or subdivisions of professional specialization to the percentage gain in the labor force of all immigrants.

more human capital will have the highest probabilities of entering high-level sectors. The following subsection examines who (which groups) is (are) most likely to make the transition into apex occupations.

Variations by Immigrant Origins

Not all groups are making the same progress. This is of course a story that has been told before in slightly different formats (Clark, 1998), but in most instances the previous presentations have not delved very deeply into fine breakdowns of the foreign born in various occupations. The graphs here emphasize the remarkable role of some groups of the foreign born in the immigrant professional workforce (Figure 4.8). Most notable are the proportions of East Asian (Indian) immigrants, among whom more than half of their employed population is in professional occupations. China, Canada, the Middle East, and the Philippines are not too different. Foreign-born engineers and computer scientists dominate some areas, computer hardware and software development, for example. The graphs confirm that it is those immigrants from origins which provide high-level education that are most likely to have their members in the professions. The initial education and training generated the human capital to make entry into the U.S. professional workforce possible. Training in India, Canada, and the Philippines, where English is the language of the universities and technical institutes, can be translated into professional jobs anywhere in the global economy. The impact of China is notable, as high-level education systems provide well-trained engineers and scientists who have gained additional training in the United States.

The change over time is also revealing. All groups have increased their presence in the professional occupations (Figure 4.8). In 1980, East Asian and Middle Eastern immigrants had about a third of their labor force in professional occupations, and the increase is remarkable. Recall that these are foreign-born professional workers, which means that a very large proportion of the most recent immigrants from India, Iran, Israel, and Japan, for example, are entering professional occupations when they arrive.

Many Indian, Chinese, Taiwanese, and other Asian workers are programmers, web experts, video consultants, and managers of new companies, or are employed in the burgeoning hardware and software businesses in technology centers like Silicon Valley, Austin, Texas, and Boston. The opportunities in the myriad new high-tech centers have been seized by the new immigrants to enter professional occupations, and by exten-

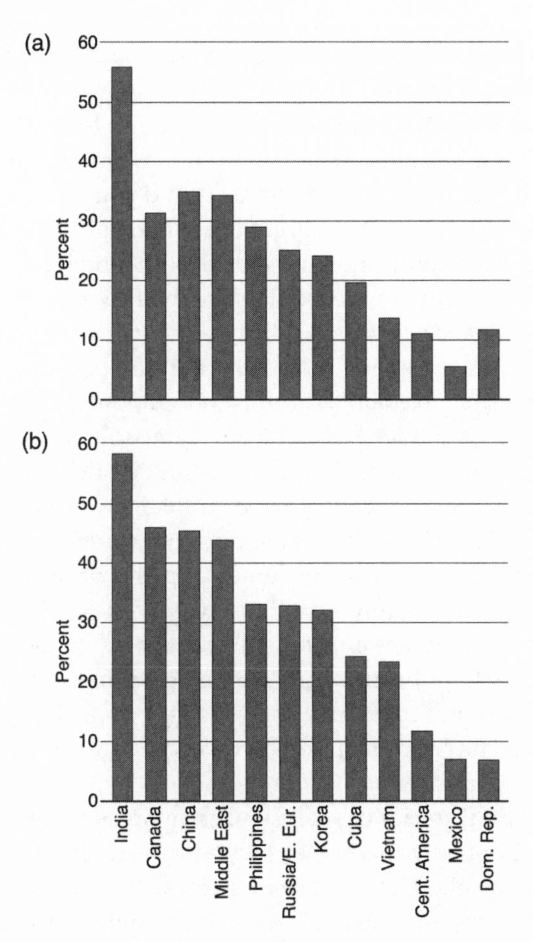

FIGURE 4.8. Ethnic-origin composition of the immigrant professional workforce by origin in (a) 1980 and (b) 2000. *Source:* U.S. Bureau of the Census, Public Use Microdata Sample, 1980 and 1990, and Current Population Survey, 2000.

sion to acquire the basis of the middle-class lifestyle. These new foreign-born professionals arrived in the United States with professional specializations and have made the most of their human capital. In fact, as we saw in Chapter 1, many of the professionals are recruited in India and other Asian countries that are known to have good education systems and a supply of high-tech professionals who want to work in the United States.[4] Whatever the process, the outcome is clear: large numbers of new members of the professions, and significant gains in human capital for the United States as a whole.

Although the story in this chapter is not about self-employed entrepreneurship per se, where a move up the economic scale is not necessarily a change in status, it would be shortsighted not to recognize the power of entrepreneurial behavior in creating the new immigrant middle class. A detailed study of foreign-born scientists and computer analysts in Silicon Valley has shown just how great the impact of the foreign born has been on the U.S. economy and on the lives of the foreign-born population in that rarefied information technology hotbed (Saxenian, 1999). The research documents the gains to the Silicon Valley economy from the highly skilled immigrants who have settled there in the past decade and a half. The special focus of Saxenian's study is the Chinese and Indian computer scientists, many of whom have created companies that are household names in the United States. Saxenian estimates that in 1998 the firms started by these immigrants generated in total about $17 billion in sales and more than 58,000 jobs. Within the findings are some important observations for our discussion of movement up the professional hierarchy. The new entrepreneurs rely on ethnic networks to create and expand their workforce, but at the same time they reach out to the mainstream information technology businesses. They employ many coethnics and so give them a chance to move up, and they rely on a stream of new well-trained immigrants from their homelands who in turn enrich the professional immigrant numbers.

The individual stories of successful companies created by immigrant engineers are impressive. Whether the gains were through stock options at industry giants like Microsoft or through starting and managing small software companies, the gains for individuals in this specialized industry were large and have had significant spin-off effects, even in the current economic downturn (Kotkin, 1999). But even though the stories of high-flying immigrant entrepreneurs in Silicon Valley make great media presentations, the more ordinary stories of teachers, laboratory technicians, nurses, and accountants are the important stories of professional success.

In 1993 Homero Luna arrived in Dayton, Georgia, from rural Mexico to find a job in the town's poultry plant. Six years later he became the publisher of the town's rapidly growing Hispanic newspaper, *El Tiempo.* The move from low-paid chicken processor to newspaper publisher in a few years is hailed as another Horatio Alger story (Tobar, 1999), but in fact it is more common than not for the most persistent and hardworking immigrants. Sometimes family resources help too. In Los Angeles, the garment, jewelry and textile-manufacturing industries have a major presence of Middle Eastern immigrants (Allen and Turner, 1997). A careful look at who the immigrants are in the entrepreneurial sectors reveals that they are often well educated and have family backgrounds which are professional. Often, too, as a new entrepreneur remarked of his background and the background of other similar immigrants, "we were the doctors, the merchants . . . we worked well with the Turks, the British and the old monarchy" (*Los Angeles Times,* September 12, 1999, p. B7). Adaptability, flexibility, and opportunism are a combination that served immigrants well in the past and continues to do so in the current expansion of the foreign-born workforce.

At the same time we have to recognize that the small group of highly educated immigrants at the top is balanced by the large group at the bottom end of educational attainment. An additional point, which is well illustrated in the foregoing discussion, is that the group at the apex of the occupational ladder is often concentrated in a few places (e.g., Silicon Valley and Austin, Texas), where there are universities and a conglomeration of technical employers. So the well-educated and the well-trained immigrants end up geographically concentrated, in perhaps two dozen locations in the United States. This geographic concentration works to the advantage of these localities with large numbers of high-earning professionals. They pay more taxes, spend more for services, and fuel housing market growth. However, the story of success in rural Georgia reminds us that the patterns are changing and that professional advance does not occur only in university towns.

The concentration of highly skilled immigrants shows up in the average levels of education. In Massachusetts, for example, a third of the immigrants who arrived since 1990 have at least a bachelor's degree and many have graduate degrees (Massachusetts Insight, 1999). Nationally, about a quarter of all recent immigrants, those arriving in the past decade, have college degrees. The very fact that a very large proportion of the new immigrants arrive with the human capital to make immediate entry into professions is often lost in the discussions of low-skilled immigrants in low-paying jobs.

Reports and discussions of the biotechnology industry, of computer engineering, and of software development often focus on the influx of Asian-born—especially Indian-born—professionals. Yet, there is also a significant flow of professionals from Africa, consisting not only of engineers and computer experts but doctors, teachers, lawyers, and business executives as well. Again, their shared common dominator is advanced education (Carrington and Detragiache, 1998). There are estimates that as many as 100,000 highly educated African professionals have immigrated to the United States. It is a classic syndrome—highly educated professionals moving to the United States, attracted by the opportunities for further professional development and for higher standards of living.[5]

Interviews with foreign-born immigrants who have gravitated to the high-tech sector underscore some basic similarities. They originate from countries with highly developed educational systems, oftentimes coming to attend U.S. universities and staying on to work thereafter. The immigrants often, but not always, came from families with highly educated parents. They have the basic skills to acquire additional training and are often willing, as have immigrants in the past, to work long hours at demanding jobs. Some come with advanced training and completed further education, whereas others come with basic high school training and use the community college system to work their way up the technology ladder. Not only do the new foreign-born immigrants with high levels of education easily move into professional occupations, they often become the founders of companies like Kopin, Sitara Networks, and Open Market that in turn provide the jobs for the new immigrants.

However, not all immigrant groups fare so well in professional advancement. In general, Mexican and Central American foreign-born immigrants are less likely to be in professions. But even though the story is less sanguine for Central Americans (in fact, there were proportionately fewer professionals among Central Americans in 2000 than in 1980), the proportion of Mexicans in professional occupations grew to 6.7%. It is worth remembering, too, that this a proportion of all employed workers 16 years of age and older. The Mexican proportion thus represents a real increase in the percentage of professionals, because very large numbers of young workers are being added to the totals.

There is surprisingly little ethnic concentration within the professions as a whole, but the aggregate professional category does conceal some specific patterns in specific occupations. Immigrants from India and the Philippines make up 33% of the medical professionals, and those from China, India, and the former Soviet Union are 28% of the "other" professional group (obviously the programmers and engineers of the high-tech sectors). Still, the wide range of ethnic backgrounds which are part of the professional workforce structure is a mark of the broad-based penetration of immigrants into the professional workforce structure.

Within immigrant origins there are concentrations, but again the overall picture is notable for the general pattern of penetration of all professions by the new immigrants (Table 4.5). Mexican and Cuban immigrants are more likely to be managers, and Russian and Eastern European immigrants are more likely to be in "other" professional activities. Immigrants from the Dominican Republic and Canada are more likely to be employed in the field of education than are other immigrant groups, a reflection of two different forces: the English-language ability of Canadians, and the large number of Dominican teachers needed for schools in New York and New Jersey. At the same time, teaching is a way of moving up and increasing occupational prestige, and many new immigrants find employment in teaching.

REGIONAL VARIATIONS

The large numbers of the foreign born arriving in the United States with skills and education, as well as those who are able to translate their previous skills into marketable professional occupations, define recognizable regional patterns. In several regions the absolute numbers and proportional gains in professional occupations are larger than they are for the United States as a whole (Figure 4.9). There are more than 1.3 million immigrants who are in professional occupations in California, and the numbers have nearly doubled in the last decade. There are an additional three-quarters of a million immigrant professionals in New York/New Jersey and another third of a million in Texas/Arizona/New Mexico and

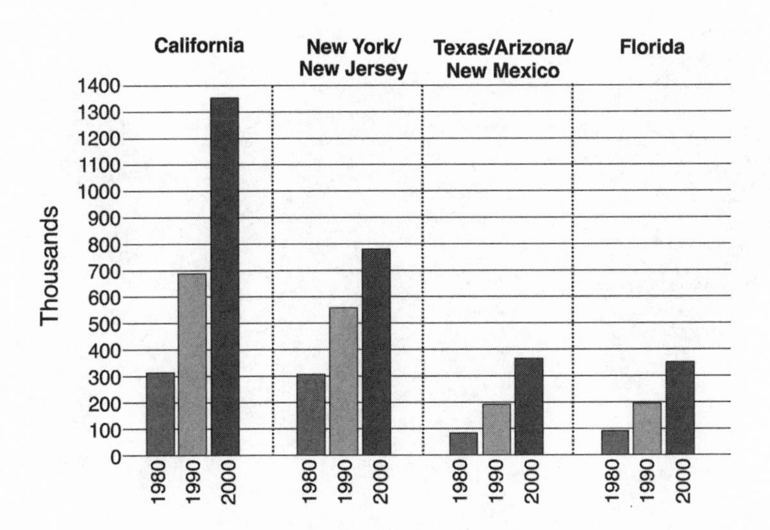

FIGURE 4.9. The immigrant professional workforce, 1980–2000. *Source:* U.S. Bureau of the Census, Public Use Microdata Sample, 1980 and 1990, and Current Population Survey, 2000.

in Florida (Figure 4.9). The rate of increase is equally notable. Of course California stands out in sheer size of the professional workforce, but the increases in Texas and the nearby Southwestern states and the increases in Florida are proportionately greater. There is no question that the immigrant flows are hardly just the poor and huddled masses so often identified in the flows early in the century, nor are they without human capital.

While there are real concentrations of professionals in the high-immigrant-population states, there are foreign-born professionals across the United States (Figure 4.10). Nationally, about 10% of professional workers are foreign born, and half a dozen states exceed this proportion. Notably and expectedly, there is a correlation between foreign-born professionals and foreign-born immigrants with middle-class status. A very large number of states have between 3 and 6% of the professional workforce foreign born, and this will likely grow. Only in the most Northern states and in some areas of the South are foreign-born professionals a small proportion of the professional workforce.

This book has continued to emphasize that the process of immigrant entry is related to educational background, but it is also related to the opportunity structure in different states. We can reliably speculate that the greater opportunities for individuals to manage small ethnic businesses may well boost the proportion of managers in California. In terms of pro-

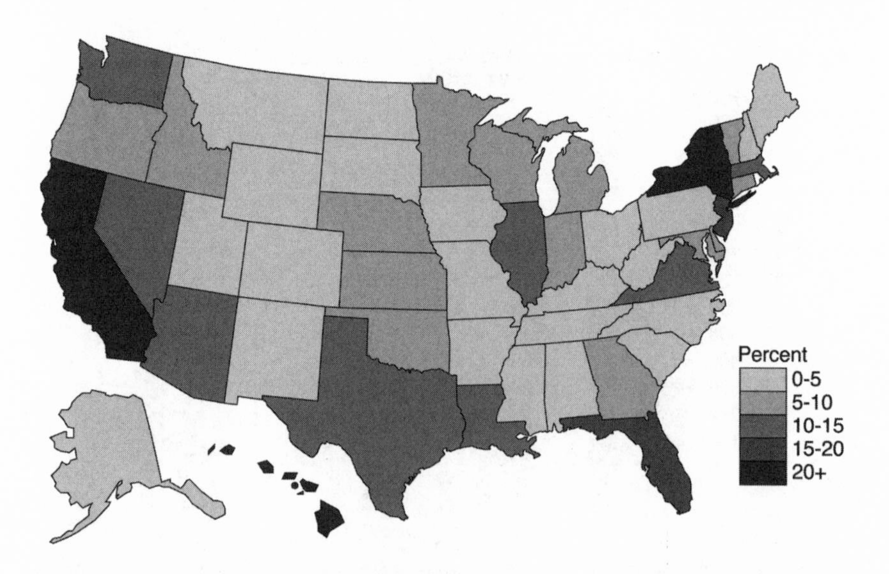

FIGURE 4.10. Percentage of foreign-born workers in U.S. professional occupations. *Source:* U.S. Bureau of the Census, Current Population Survey, 2000.

portional gains, now more than a quarter of the professional workforce in California is foreign born, and it approaches 20% in New York/New Jersey and Florida. It is still lower in Texas but close to the national average (Table 4.6). For specific professional occupations, the overall patterns across the large-immigrant-impact states are similar, though there are subtle differences especially between Florida and California and Texas/Arizona/New Mexico (Figure 4.11). California and Florida have quite similar patterns of professional gains across all the professional categories. For example, managers, medical workers, and other professional activities are all about the same level in 2000, although there are differences in the rate of growth across the different professional specialties. In addition, in both these states, there are about the same proportions of lawyers as of those in education. New York, while showing a general pattern which is like that of California and Florida, is notable for the recent slow growth of the medical specialty professions. However, like California and Florida, a quarter of all professional-medical positions are now filled by immigrants. Texas, Arizona, and New Mexico have lower levels of foreign-born professional proportions and the only instance (in the legal professions) of a stable relationship in the past decade.

PROFESSIONAL PATTERNS BY ETHNICITY AND GENDER

Some of the differences in aggregate professional advance are related to immigrant composition. The overall review earlier in this chapter established that Asian (apart from the refugee populations from Southeast Asia) and European and Middle Eastern populations are more likely to be

TABLE 4.6. Immigrants as a Proportion of the Professional Workforce

	1980	1990	2000
United States	5.9	8.6	11.2
California	11.6	17.6	25.7
New York/New Jersey	11.3	15.4	19.2
Texas/Arizona/New Mexico	4.6	7.6	9.5
Florida	10.1	13.9	18.2

Source: U.S. Bureau of the Census, Public Use Microdata Sample, 1980 and 1990, and Current Population Survey, 2000.

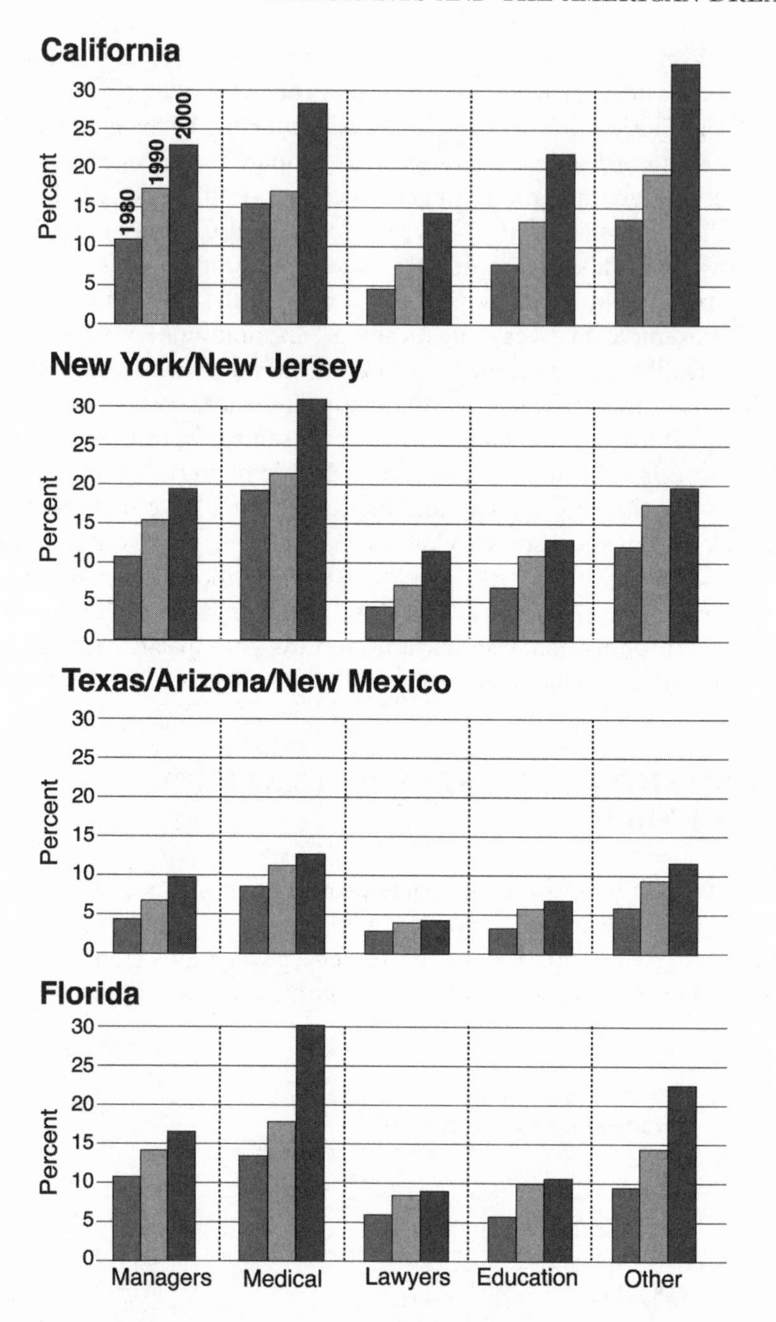

FIGURE 4.11. Immigrant proportions of the professions for large-immigrant-impact states, 1980–2000. *Source:* U.S. Bureau of the Census, Public Use Microdata Sample, 1980 and 1990, and Current Population Survey, 2000.

in professional occupations, and that Mexican and Central American populations are less likely to be in those occupations. Thus, the pattern of relatively greater proportions of Mexican and Central American immigrants in Texas and the Southwest reduces the overall proportion of the foreign born who are likely to be in professional specializations. At the same time, immigrants in the professional occupations are high in California, where there is a large Mexican immigrant population, so it is clearly a complex process with varying outcomes.

The composition of the foreign-born professional workforce is notable for its diversity (Figure 4.12). While Asians are a large proportion of the total foreign-born professional workforce in California and New York/New Jersey, the wide range of other "ethnic" origins varies from 20 to 40%. It is clear that a wide range of immigrants are making it to apex occupations and providing an enriched professional class in the high-

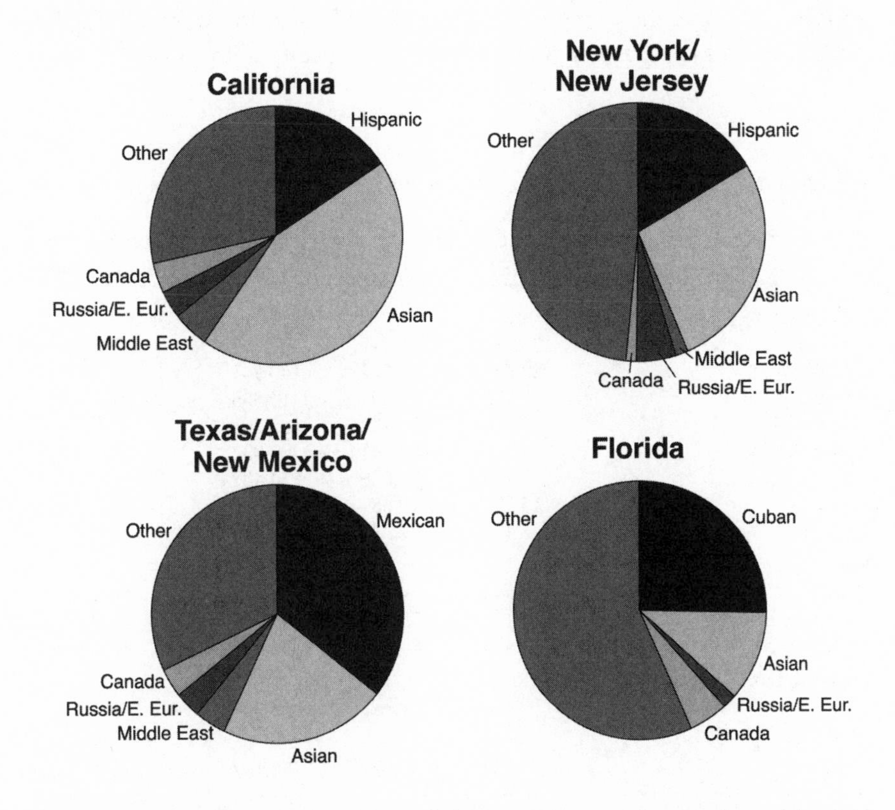

FIGURE 4.12. Ethnic composition of the immigrant professional workforce in 2000. *Source:* U.S. Bureau of the Census, Current Population Survey, 2000.

immigration regions of the United States. While on average Asian and Middle Eastern groups were more likely to have moved into the professions than were Hispanic groups, are there success stories within the Mexican immigrant community?

The answer to that question is clearly "yes." However, to uncover these findings requires a consideration of the total numbers, and not just of the proportions of groups of the immigrant population who enter professional occupations. The regions with the largest Mexican populations, California and Texas/Arizona/New Mexico, have had very rapid increases in the Mexican professional population. California had less than 30,000 Mexican professional workers in 1980, but by 2000 the number was 150,969 (Table 4.7). Similarly in Texas, the increase in professional Mexicans immigrants was from 16,960 to almost 100,000. Clearly, large numbers of the Mexican foreign born are making the transition to professional occupations. The successes for now are greater in the large-immigrant-impact states, but the process will certainly spill over to other geographic locations. And the point is that even for immigrants with lower levels of

TABLE 4.7. Total Number of Professionals by State and Selected Countries of Origin with More Than 50,000 Persons

	1980	1990	2000	% Professionals
California				
Mexico	30,780	86,315	150,969	7.3
Philippines	28,560	80,072	140,941	31.1
India	8,720	23,594	115,555	71.5
China	27,740	74,212	107,702	36.7
Middle East	8,540	43,841	82,932	38.7
C. America	7,180	27,038	54,482	11.9
Russia/E. Eur.	15,100	20,659	50,504	44.2
Canada	27,440	32,335	50,356	56.7
New York/New Jersey				
India	18,040	34,270	79,119	50.6
Russia/E. Europe	15,100	35,133	52,988	24.8
Philippines	14,800	30,580	50,430	46.5
Texas/Arizona/New Mexico				
Mexico	16,960	11,365	98,767	8.2
Florida				
Cuba	33,160	55,829	58,187	18.2

Source: U.S. Bureau of the Census, Public Use Microdata Sample, 1980 and 1990 and Current Population Survey, 2000.

human capital, it is still true that new immigrants are making their way up the occupational ladder. The fact that there are about as many Mexican immigrant professionals in the United States (316,528) as there are those from China (357,518) and India (370,059) emphasizes that even though the Asian and Middle Eastern groups are doing proportionately extremely well, a function of the high levels of social and human capital and previous training, there is a pattern of success which extends to immigrants who are often seen as educationally less well equipped to penetrate apex occupations.

Outcomes for Women

Women, foreign born and native born alike, play a much greater role in the workforce than they did a generation ago. In the past two decades, native-born women have increased their workforce participation rate from somewhat more than 40% to nearly 50%. Foreign-born women have also increased their participation rates, but they have not achieved such high levels of participation and recently have been relatively stable in their participation rates at around 42% of the total immigrant workforce (Table 4.8).

The story of occupational success is more complicated for women. In general their participation in the professions does not match their availability or presence in the population (Figure 4.13). Only in the medical professions do women exceed their general proportion in the population. In all the other professions they have increased their role over time, but even though these increases are impressive they have some distance to go to achieve parity. As for professional categories increasingly dominated by women, some are dominated by foreign-born women to an even greater extent than they are by native-born women. For example, foreign-born women dominate the same parts of the medical (especially nursing) and educational professions (Table 4.9). For the United States as a whole

TABLE 4.8. Women in the Professional Workforce

	1980	1990	2000
Total	467,560	1,102,793	926,415
Percentage of immigrants	35.2	42.0	42.7

Source: U.S. Bureau of the Census, Public Use Microdata Sample, 1980 and 1990, and Current Population Survey, 2000.

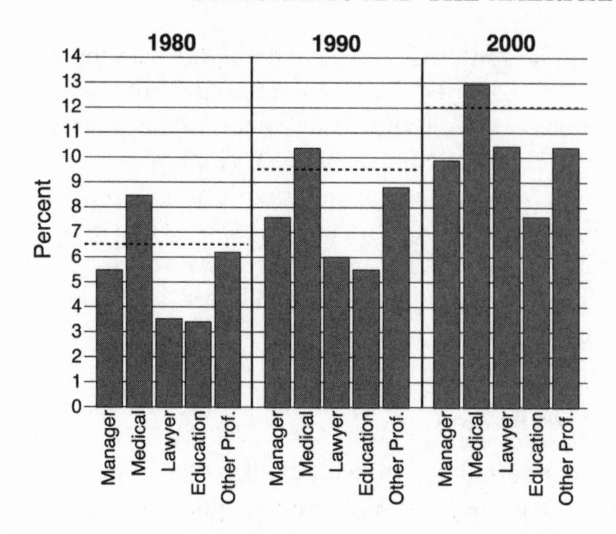

FIGURE 4.13. Immigrant women as a percentage of the female professional workforce in the United States (dashed line is the percentage of the total women who are immigrants). *Source:* U.S. Bureau of the Census, Public Use Microdata Sample, 1980 and 1990, and Current Population Survey, 2000.

TABLE 4.9. Percentage of Immigrants in Professional Occupations in 2000 Who Are Women

	Managers	Medical	Education	Other
United States	42.3	62.9	62.2	27.6
California	41.9	65.9	73.2	23.8
New York/New Jersey	41.2	74.6	62.8	30.9
Texas/Arizona/New Mexico	42.3	62.3	76.8	31.1
Florida	36.8	59.4	70.4	22.0

Source: U.S. Bureau of the Census, Current Population Survey, 2000.

they make up 63% of the immigrant medical workforce and a similar proportion for the immigrant education workforce.

Anecdotal evidence provides a window on the success stories of the immigrant women who have sufficient education and skills to enter the service professions, especially those in the medical and educational fields. Young women in the Philippines saw opportunities in the service professions, especially nursing, and used the greater ease of migration to tap into the possibilities of upward social mobility by moving to the United States. Now, the clinical and laboratory technicians are increasingly Filipinas, as in the nursing profession as a whole. This trend is a natural outcome of the shortage of nurses in the United States in general. However, it is also an outcome of the changing status of nursing, and the changes in the medical service profession in general. There has been a shift to create a flexible workforce in the service professions, as in manufacturing and other parts of the economy. Nurses are often hired as contract labor and are no longer affiliated with the hospitals who employ them. Along with this change in the workforce, it is possible that being a nurse is no longer as prestigious as it once was. Nonetheless, the opportunities in these professions are still considerably more attractive than those in janitorial and other more menial service jobs.

The role of women in the professions for the large-immigrant-impact states shows just how important the contribution of immigrant women is to the overall workforce (Table 4.9). In most professions the proportion of women has increased over time, and now in California 73% of the immigrant workforce in education are women and in New York/New Jersey about 75% of the immigrant medical workforce are women. Women dominate the immigrant workforce in the medical and educational fields in all states as they do for the United States as a whole. Most notable is the proportion in managerial categories. In California, in Texas/Arizona/New Mexico, and in Florida, women are now more than 40% of the immigrant manager workforce. They are managing small businesses, retail outlets, and small service companies. Their overall participation as managers has increased by some 10–15% in the two-decade period we have been examining. The data suggest that of the new immigrants arriving and entering the workforce, the women are narrowing the gap with men: they are nearly equally likely to be in managerial positions, and they are much more likely to be in the educational and medical professions. Women still lag in the other professions—in engineering, in public relations, and computer technology—in the areas with the greatest educational requirements.

Still, immigrant women have not reached parity with the female workforce in general. In some states and in some professional occupations, they are at or above the level of employment of native-born women, but in many cases the progress is slow and they are losing ground relatively (Figure 4.14). So many native-born women have moved into the U.S. labor force in the last two decades that the relative proportion held by foreign-born women has shown slower growth. In some cases there has been a slight decline, as in the case of lawyers in New York/New Jersey, for example.

Other evidence from studies of labor market participation also suggests that women have a mixed experience in the U.S. labor market (Schoeni, 1998a). There are quite different levels of labor force participation across immigrant women. Schoeni (1998b) finds that immigrant women from Japan, Korea, and China have the greatest levels of participation, and we have seen that they also have the highest levels of participation in the professional occupations. Schoeni's evidence revealed, as the study here shows for the professions, that Filipino women are distinctive in their rates of labor-force participation. Schoeni (1998b) found that 80% of the prime-age Filipinas were working in 1990. As for immigrants as a whole, the relatively high levels of education of Asian women explain the gap between the participation of many Asian women and Mexican women.

They have been successful in some areas, less so in others. But again, for women as for men, human capital, previous education in particular, is the crux of explaining why some do well and others do less well. The gains in the professional occupations are closely tied to years of schooling and English-language skills (Schoeni, 1998a).

SUMMARY AND OBSERVATIONS

The findings in this chapter substantiate findings from other studies and provide a more detailed understanding of what is happening to the professional occupations in the United States and the large entry-point regions. The data and analysis in this chapter confirm the impressive upward mobility immigrants can experience in the United States. Socioeconomic status has increased for many, and with status comes at least reasonable incomes. Proportionately, more immigrants are in the middle class than they were two decades ago. That status is represented by large numbers of new immigrants who have moved into middle-class occupa-

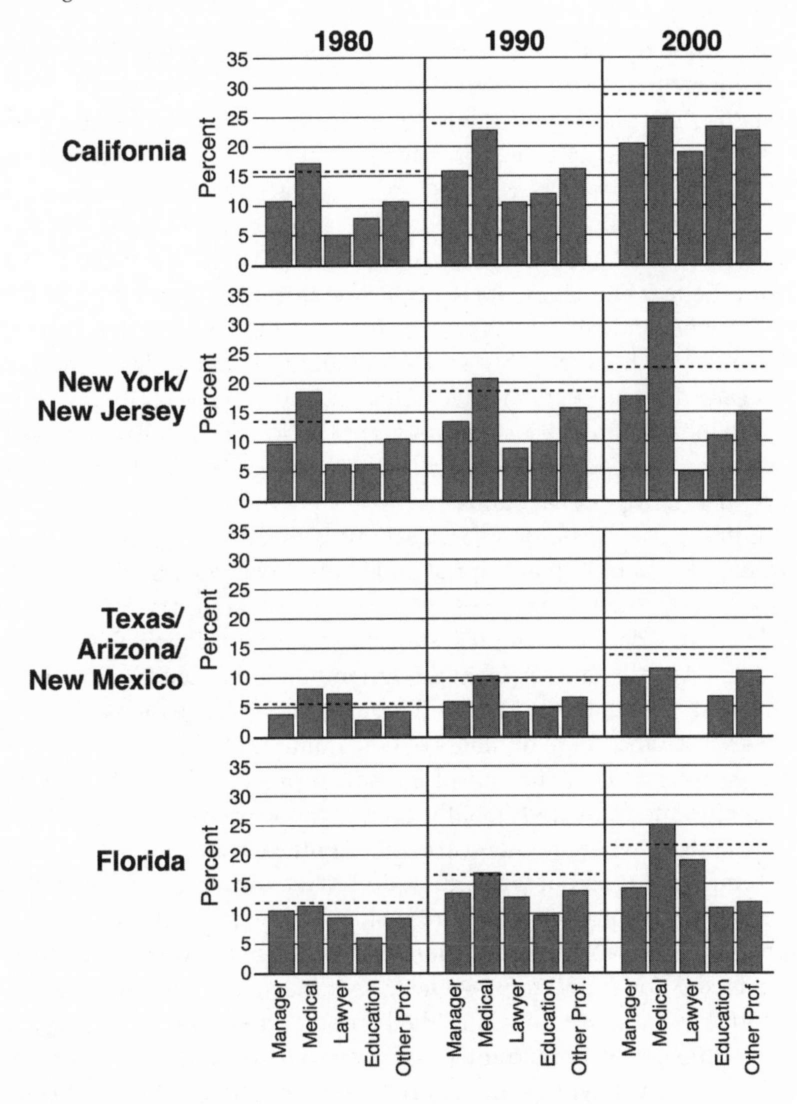

FIGURE 4.14. Immigrant women as a percentage of the female professional workforce by state (dashed line is the percentage of the total women who are immigrants). *Source:* U.S. Bureau of the Census, Public Use Microdata Sample, 1980 and 1990, and Current Population Survey, 2000.

tions: nurses, doctors, engineers, and personnel managers. Even though the proportion of the professional occupations that are occupied by immigrants has increased only modestly, from about 21% of the immigrant workforce to about 24%, that proportional increase represents a growth of nearly 3 million workers who are now in professional occupations.

Even research on undocumented immigrants shows that there is improved occupational status over time (Powers and Seltzer, 1998; Powers, Seltzer and Shi, 1998; Kraly, Powers, and Seltzer, 1998). The studies of Powers and colleagues tend to focus on the lower half of the occupational status hierarchy, but still the evidence is fairly clear that immigrant earnings and occupational status improved between their first jobs and the jobs they held just prior to application for legalization. Immigrants are experiencing upward mobility, and legalization has helped advance their occupational status.

Entry into professional occupations still favors those immigrants with high levels of human capital and skills. Consequently, the gains are greatest for the new immigrants who come from countries with good or excellent secondary education systems: India, Japan, China, South Korea, and some African nations. Even though immigrants from Mexico are represented at a much lower level in the professional occupations, this chapter has shown that large numbers of new immigrants from Mexico are becoming professionals. The gains for a new professional middle class exist for immigrants from traditionally poorer countries as they do for those from countries where human capital is significantly greater.

Immigrant women, when they do have education and skills, make up a significant proportion of professional activities that are traditional strongholds for women: education and nursing. However, the story is more mixed for immigrant women, and they are much less likely to be managers or professionals in the computer and high-technology sectors. To this extent immigrant women are not dissimilar to native-born women. However, while native-born women on the whole have been increasing their participation in the labor force, immigrant women's participation rates have been stable or sometimes losing ground.

The analysis in the book to this point has presented a strong albeit complex story of immigrant success, and the increase in the number in the professions is certainly a part of that story. The story is more complex than that about previous flows, because at least some of the new immigrants are arriving with high levels of education and professions. They are middle class in many ways already. The old notion of immigrant success over one, two, or more generations needs to be tempered with the realization that significant numbers of immigrants are arriving with substantial socioeconomic status. At the same time, will the immigrants who are coming with fewer skills be able to make the transition to more successful lives, and will their children go on to college and join the professions? That story is less clear, and the increasing numbers of low-skilled migrants may make the path more difficult for new arrivals than it was for earlier arrivals. Only time will answer these questions.

NOTES

1. Most recent work on immigrants and labor markets has focused on the wage gap between the native born and the foreign born, and on the effects of new immigrants on the wages of the native born. These are contentious issues although a consensus seems to be that the wage gap between the native born and the foreign born is increasing, because the native born are doing well and the effects of low-cost labor is mostly on other recent arrivals in low-wage jobs. The question posed in this chapter is focused on the labor market experiences of the new immigrants.
2. These are the modern equivalents of earlier waves of Irish police and firefighters, German brewers, and Italian cafe owners.
3. The growth of the foreign-born workforce was nearly 200% in total. The growth for foreign-born men was a little larger than that for foreign-born women, but both grew rapidly in the 1980s and 1990s. Native-born women increased their labor force participation significantly: they added an additional 16 million workers in the period 1980–2000. In the same period, foreign-born women in the workforce increased by 5 million.
4. There is a vigorous debate about whether the new IT (information technology) foreign professionals are taking jobs from native-born programmers and whether the IT industry is unwilling to hire and/or retrain older programmers (Reports to the U.S. House Judiciary Committee, Subcommittee on Immigration, April, 1998).
5. There is an ancillary question about whether this "brain drain" is creating a barrier to internal economic growth in the African countries. While the U.S. gains, African countries are losing the gains of their educational infrastructure to the United States and Europe.
6. A reminder: throughout this chapter, as in Chapter 3 and in the chapters that follow, I use data on California and Florida and on New York/New Jersey and Texas/Arizona/New Mexico as geographic aggregations, but I sometimes use the first state as shorthand for the grouping.

CHAPTER 5

★ ★ ★

Reaching for Homeownership

Homeownership is an integral part of the American Dream and our story of progress to the middle class, and nowhere is the story more apparent than in the surnames of recent homebuyers in the Los Angeles metropolitan region. While the "Smiths" and "Johnsons" are still among the biggest home purchasers in the United States as a whole, in Los Angeles the Smiths are only ninth in the ranking and the Garcia, Martinez, and Hernandez households are all much more likely to be the homebuyers (Table 5.1). There is a fundamental change occurring in the homebuying industry in California, the state with the largest immigrant population, a change which is likely a harbinger of things to come in the United States as a whole. The new immigrants are buying homes, accumulating assets, and putting down roots.

As immigrants arrive in the changing housing markets of the global gateway cities they often begin their lives in inexpensive rental housing, but—as shown in the analysis below—these immigrants in a few short years are successfully penetrating the housing markets in these cities and becoming homeowners.[1] It is true that the increased demand for housing, driven by the increased immigration, sometimes creates problems of affordability and access, but it is ownership which increases the affiliations than unify people at local scales and bind them—however loosely—to the host society. As stakeholders in the status quo, homeowners are

TABLE 5.1. Most Common Last Names of Recent Homebuyers in the United States and California

United States	California	Los Angeles
Smith	Garcia	Garcia
Johnson	Smith	Lee
Williams	Lee	Rodriguez
Brown	Johnson	Lopez
Jones	Lopez	Gonzalez
Miller	Martinez	Martinez
Davis	Hernandez	Hernandez
Anderson	Rodriguez	Kim
Wilson	Nguyen	Smith
Rodriguez	Gonzales	Perez

Source: Dataquick, March 2001; *www.dqnews.com*, and *Wall Street Journal*, September 10, 1998.

integrated into local communities, whatever their origins or recency of arrival.

Thus, homeownership is connected not only to wealth accumulation but also to putting down roots, and by extension to social integration, if not assimilation. Putting down roots, as part of the process of homeownership, may be the least tangible quality of that process, but community participation and involvement may be the most important element in the integration process. Buying a house is purchasing a specific location, and with the location a commitment to a particular community and its services. Whether the neighborhood is homogeneous or in transition, by owning a part of the community the purchaser is in fact making a commitment to society as a whole. Several studies support the idea that homeownership encourages active community participation (e.g., Lee, 1999).

Homeownership has expanded significantly in the past three or four decades in the United States, part of the long history of federal government support of increasing homeownership nationwide. Middle- and upper-income families, in particular, have been able to take advantage of generous tax deductions, and federal and state governments have promoted ownership rates through loan guarantees and low-cost mortgage insurance. Particular government programs like those for veterans after World War II greatly expanded homeownership across a wide section of moderate-income households. The justification for this national policy of increasing homeownership was that it invoked wealth accumulation on the part of families and was a way to generate jobs and stimulate economic growth. Homebuilding was then—and still is—seen as an industry that has a multiplier effect far beyond housing itself.[2] All of these policies have served to increase homeownership from a little above 40% before World War II to 68% by the end of the 20th century.

It is in the context of a nation of homeowners that most contemporary immigrants arrive in the United States. There is a sharp contrast from immigrant waves of a century ago coming to cities in which the housing stock was largely rental, and of course of much lower quality overall than housing in present-day U.S. urban centers. Now immigrants are arriving in a changed housing market. We can ask how the recent immigrants are doing in the cities and neighborhoods where they are currently arriving. Are they becoming homeowners, and how quickly can they make this most basic adaptation in the United States?

HOMEOWNERSHIP IN THE UNITED STATES

Homeownership has always been a part of the American Dream of upward mobility and financial security. Ownership offered that most quintessential part of American mythology—independence and freedom, specifically freedom from the capriciousness of landlords and security against financial dependence. Of course, the dream in reality was affected by the continuing problems of affordability for low-income households and discrimination against certain ethnic groups; nevertheless, homeownership has become an ingrained part of the goal of "making it" in America.

There are sound reasons for expecting a preference for homeownership. Owning a house increases "housing security" and gives greater control to owners over their physical surroundings. It lowers real monthly costs over time, protects against unanticipated rental increases, and—perhaps most importantly—generates wealth over the long term. It is the last-named goal, wealth generation, which is an important part of the socioeconomic progress of all families, native born and foreign born alike. Even so, there is research which suggests that the major benefits of ownership go to wealthier families.[3]

The major attractions of homeownership are linked directly to asset accumulation, increases in housing quality, housing security, and community connections. The financial gains for homeowners have been considerable, and even though house prices do fluctuate the general steady increase in house prices in the last two decades has provided a significant addition to family assets. In addition to house price appreciation, owners also gain from building equity by repaying their mortgage over a 20- or 30-year period. For most families the family home is still the main asset which they have accumulated over their working lifetime. By the early 1990s housing equity represented about 45% of the net worth of the average homeowner (G. McCarthy et al., 2001). Of course the proportion of household assets represented by housing was lower for white households, who are likely to have other investments, and higher for ethnic and immigrant households. For Hispanics, for example, housing equity represented nearly 61% of household net worth (G. McCarthy et al., 2001).

The vast majority of adults in the United States, nearly 90% of homeowners and 72% of renters, regard homeownership as a sound investment (Fannie Mae, 1995; National Association of Realtors, 1992), and justifiably so.[4] Because of the forced savings and the acquisition of equity, homeowners have greater access to capital than do renters. The capital might come in the form of home-equity lines of credit or the proceeds of a second mortgage. In this way the homeowner can raise funds for a wide range of needs—from college tuition to general consumer expenditures. Of course, the ability to service the increased debt burden is in the end related to a stable income, and in this sense household income is a critical component of both initial ownership and maintaining ownership through changing economic times.

Perhaps as important as the raw financial gains are the psychic feelings that ownership brings: control over the physical living space and the freedom to modify, enlarge, and beautify the living environment. In addition, the quality of residences is greater for owners. Owner-occupied units are in general larger and far less likely to have physical problems.[5] Finally, ownership often brings better neighborhoods, sometimes in the suburbs, with all the other concomitants of good schools, better community services, and lower crime rates.

Homeownership and Assimilation

Homeownership is linked closely to assimilation, at least as measured by English-language use. Those who had greater English proficiency were much more likely to be homeowners (Alba and Logan, 1992; Krivo, 1995). We know that assimilation is more than using English, but it is an indicator of social integration and adaptation.[6] It is true that it takes time for immigrants to become owners; the effect of foreign birth does not disappear for as long as 30 years. At the same time one can argue that 30 years is about what we would expect for the process of assimilation.

It is important to reiterate that assimilation is a process that occurs over very long periods of time; it does not occur in a decade or two, and full social integration is much more than simply owning a home. At the same time, as I argued earlier in the book, homeownership is one of the steps of assimilation, of becoming part of the larger community by buying into communities and neighborhoods, and being a part of the daily activities of local communities. In essence, assimilation is simply a code word for the process whereby successive generations become part of a new society.

Becoming a homeowner is important because housing and its location are central determinants of assimilation in the sense that they structure access to quality schools and employment opportunities, and affect the exposure to all of the externalities which come with particular neighborhoods. This research emphasizes that those who can gain access to housing are likely to also make major inroads into the new society. Moreover, ownership by first-generation immigrants is more likely to lead to ownership in succeeding generations.

We already know a good deal about homeownership and the factors which influence those attainment rates by immigrants. The causal nexus here is complex, of course, but the interpretations are supported by research which shows that ownership increases with income, with high levels of educational attainment, and with being in a professional occupation. It also increases with age, as in general aging is related to greater human capital and higher incomes. These variables increase the likelihood of ownership for immigrant and nonimmigrant households alike. Households with lower incomes are less likely to own single-family housing. Higher incomes not only lead to single-family housing consumption, they are translated into better-quality housing and newer housing. As expected, immigrant groups do not own houses at the same rate as the native born, although many immigrant groups are attaining higher levels of ownership than those of native-born African Americans and in some cases those of native-born Hispanics. Finally, the longer an immigrant is in the United States, the more likely he or she is to become a homeowner.

Wealth effects are also thought to be important in the ownership process but are much harder to document (Bourassa, 1994; Myers, Megbolugbe, and Lee, 1998). However, there is evidence that wealth and homeownership are closely interrelated. In fact, they may be mutually determined, each increasing in anticipation of the other (Haurin,

Hendershott, and Wachter, 1996, 1997). Those who want to own their own homes increase their earnings effort, and homeowners gain in wealth from owning a house. This is an important part of the way in which immigrants do—and will—participate in assimilation to higher socioeconomic status. In turn, occupational achievements are translated into homeownership too (Myers and Park, 1999). In a sense, occupational achievement can be seen as preceding earnings. In this conceptualization, increased occupational achievement leads to greater earnings, which can be translated into higher homeownership. Of course, occupational attainment is in turn closely related to educational level.

TRAJECTORIES TO HOMEOWNERSHIP

Immigrants become owners as they age and acquire wealth. It is the same process for immigrants and the native born alike. The association between the obvious economic factors facilitating homeownership and the actual levels of homeownership attained over time are best seen as a trajectory in which a group (a cohort) of the population advances into ownership as they age. Research by Myers and his colleagues examines the level of ownership of particular cohorts at successive points in time and computes each cohort's progress across the decades (Myers and Lee, 1998). For example, the cohort of immigrants aged 25–34 in 1980 is nearly equivalent to the cohort of immigrants aged 35–44 in 1990 and the cohort aged 45–54 in 2000. Calculating homeownership rates for each age group at each point in time approximates the trajectory of the initial 25- to 34-year-old cohort and their pace of progress. The aforementioned researchers find that Asians make more rapid progress than Hispanics, consistent with higher rates of naturalization (Clark, 1998) and greater facility in English (D. Lopez, 1996).

The trajectories of households over time furnish a revealing perspective on the temporal process of access to homeownership. The detailed Public Use Microdata Sample (PUMS) data from the U.S. Censuses for 1980 and 1990[7] provide the means by which the ownership rates can be calculated for immigrants in 1980 by 5-year age groups (cohorts): 15–19, 20–24, and so on. The data from the PUMS for 1990 can be used to calculate the rates of homeownership for immigrants who were already here in 1980, by the corresponding age groupings in 1990 (25–29, 30–34, etc.). It is then a simple matter to diagram the change in ownership for a 1980 age cohort 10 years later. The outcome is a "trajectory" graph of changes between 1980 and 1990.[8] The graphs for the U.S. native born show a rapid increase in ownership as young cohorts rapidly become homeowners (Figure 5.1). In 10 years the youngest age cohorts triple and double their rates of ownership. As each cohort ages, its trajectory becomes less steep. Older households are not, in general, buying homes; they are transitioning to retirement complexes and perhaps to rentals. The foreign born trace the same path, albeit at a slightly lower trajectory.

The same technique can be used to plot the paths of the foreign born in total, and for the largest ethnic groups for each region (Figure 5.2). The trajectories are similar for all foreign born across the states, but at the same time there are meaningful differences among individual ethnic groups. Ownership attainment rates for the 1980 arrivals reach 70% at the peak years of ownership for all immigrants except those in New York/ New Jersey, and rates are higher for specific groups. For each state, the average for all immigrants is compared with a very-successful and a less-successful immigrant group.

In California rates peak at nearly 72%, but for immigrants from the

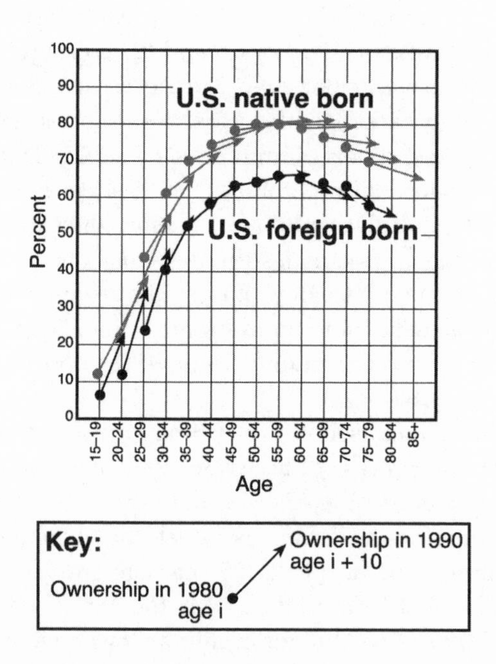

FIGURE 5.1. Cohort trajectories in 1980–1990 for native born and foreign-born U.S. households. *Source:* U.S. Bureau of the Census, Public Use Microdata Sample, 1980 and 1990.

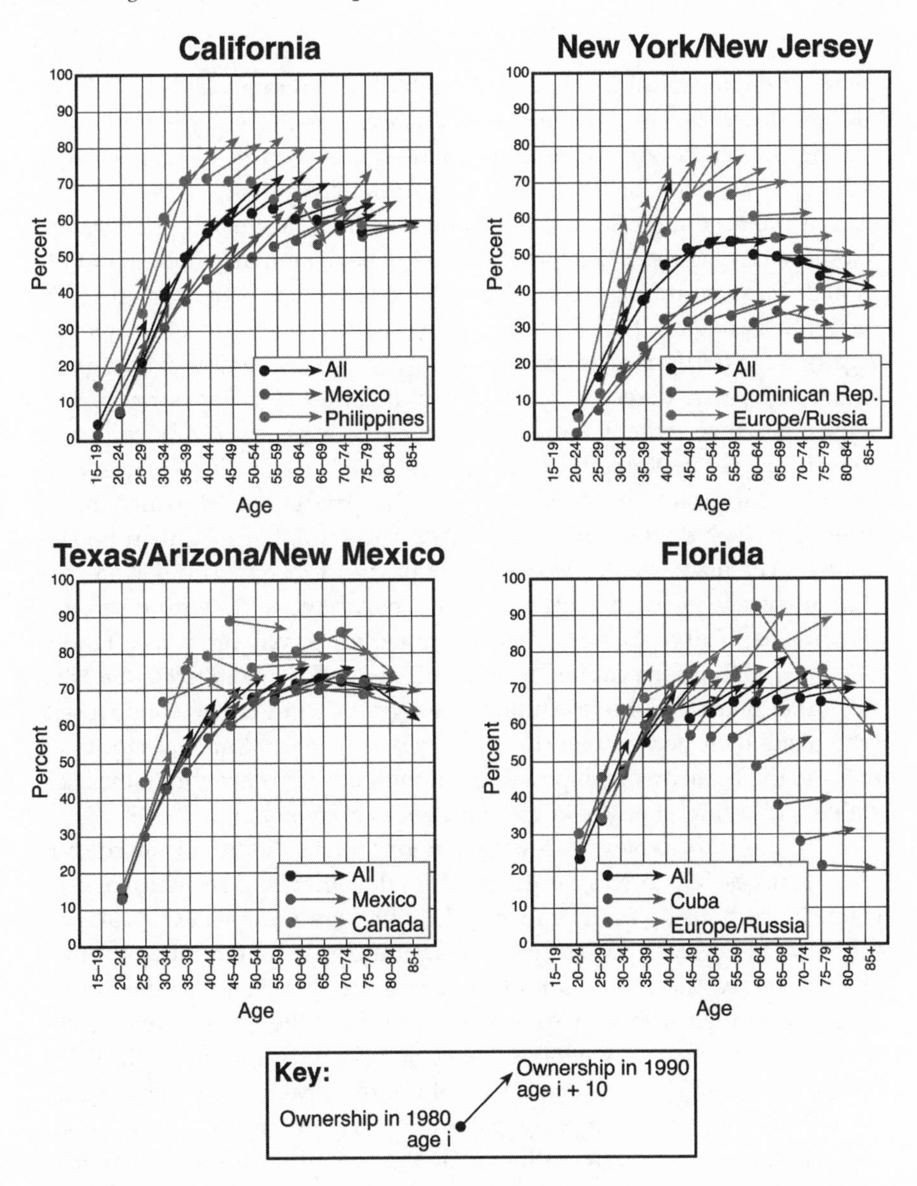

FIGURE 5.2. Cohort trajectories 1980–1990 for immigrant households. *Source:* U.S. Bureau of the Census, Public Use Microdata Sample, 1980 and 1990.

Philippines the rates reach 82%, and this happens at an earlier age than for the foreign born as a whole. Although Mexican immigrants to California have lower levels of ownership at the peak, about 63%, it is worth reiterating that these are foreign-born households who in a decade made gains in ownership of 10–12 percentage points. The pattern is the same even if the absolute rates are not as high as the rates for all immigrants. As shown in Figure 5.2, Texas and Florida have similar patterns, but although Mexican immigrants to Texas have lower levels of attainment by cohort, it is notable that the rates climb above the 70% level by age 55. Again, these findings are important evidence of considerable cohort gains in only 10 years. Rates are very high, and cohort advance is rapid. The same is nearly as true in Florida, although the increases for older Cuban cohorts are not sustained. Of course, the number of older Cubans who were in Florida in 1980 was much smaller. The falloff in ownership is quite remarkable after age 55. Clearly, a large number of Cuban households never managed to become owners, and those who did were not able to sustain the ownership rates. This may have to do with when immigrants left Cuba. Arrivals after the initial influx may have been family members without the capital of the wealthier Cubans who left Cuba after the initial phases of the revolution. Immigrants from Europe and Russia who make their home in Florida have very high and rapid trajectories of entry to the homeownership market, though the gains are sometimes unstable with rapid changes in the trajectories of ownership.[9]

The New York/New Jersey story stands out as different from the relatively successful stories of immigrants in the other homeowner markets. Gains in the housing market are smaller, although European and Russian immigrants do have trajectories that are much like immigrants in other states. Clearly, the lower cohort advances are related to poorer immigrants from the Caribbean, especially those from the Dominican Republic, who barely reach ownership levels of 40% and do not sustain those levels. However, as in the discussion of middle-class gains in New York/New Jersey, the results on participation in ownership for the foreign born will be influenced in New York City by the overall lower rates of ownership in general. Ownership rates are about one-third of the national average in New York County (Manhattan), Bronx County, and Kings County (Brooklyn). They are higher in Queens where there are many immigrant households. Still, the patterns in the diagram of progress to ownership are quite similar to patterns in the other large immigrant states.

As we might expect, ownership rates are lower for more recent entrants: they are younger and have had less time to acquire the assets to make ownership possible. Duration affects ownership (Figure 5.3). Ownership rates indeed decline for more recent entrants—or, if you will, they increase for those who have been here longer. White foreign-born ownership rates are higher than those for Asians and Hispanics, but the increase in ownership with duration is notable. Of course it is intertwined with aging, but as we have just demonstrated it is the aging (and its associated increases in wealth and assets) that makes ownership eventually possible for very large numbers of both immigrants and the native born.

A specific analysis of age cohorts and arrival times in Southern California provides a number of additional revealing findings about homeownership (Myers et al., 1998). Immigrants who arrived at particular points are studied separately from other immigrants who have been in the county longer or shorter time periods. The ownership rates for native-born whites climb steeply.[10] While almost no households with heads under 24 years of age own their homes, by age 25–34 almost 40% do. The findings show that Asian and white immigrants achieve very high levels of ownership soon after arrival and these rates often exceed those of the native born. Hispanic immigrants also make significant gains over time, although they start out at much lower levels. Even though many Hispanic households never achieve the same levels of ownership as the native born, they do sustain significant penetration into ownership

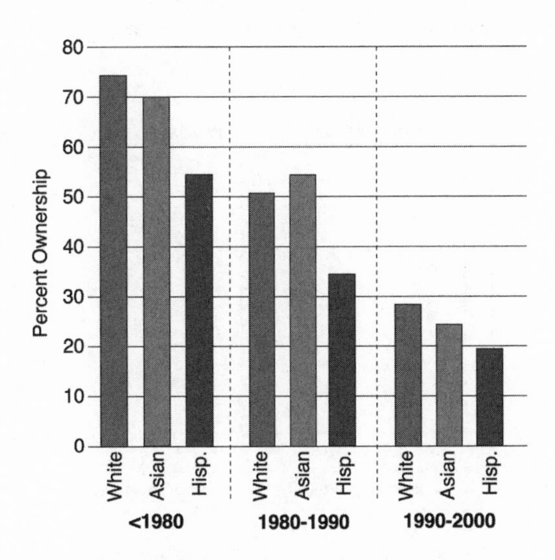

FIGURE 5.3. Homeownership rates by period of entry into the United States. *Source:* U.S. Bureau of the Census, Current Population Survey, 2000.

(Figure 5.4). Even though the data are for Southern California alone, they are useful in the sense that they compare successful access of a relatively expensive housing market.

The gaps between successive cohorts are indicative of the lag between cohorts. In the younger cohorts the 1990 groups do not reach the same levels of ownership as the 1980 same-aged cohorts. The drop in ownership rates is driven by the lower overall homeownership rates that are carried forward over time. Trajectories for foreign-born households are remarkably steep also. In addition, for earlier arrivals the rates of ownership do reach 50% for older age groups.

A further study of age cohorts and arrival times in Southern California, controlling for income, education, and marital status, provides a number of additional important findings about homeownership (Myers et al., 1998). When the trajectories from these models are plotted they reveal that income and price matter, as we would expect, and so too do the demographic composition of the cohorts. Ownership increases with higher permanent income, marriage, and higher education, and in markets with lower prices. These findings are not unexpected, but they confirm the overall gains for different entry cohorts.

The trajectories leading to homeownership are replicated with respect to housing quality. For all groups there are sharp upward trajectories of higher-value housing consumption. In particular, older birth cohorts are less likely to live in lower-value housing (Lee, 1999). These

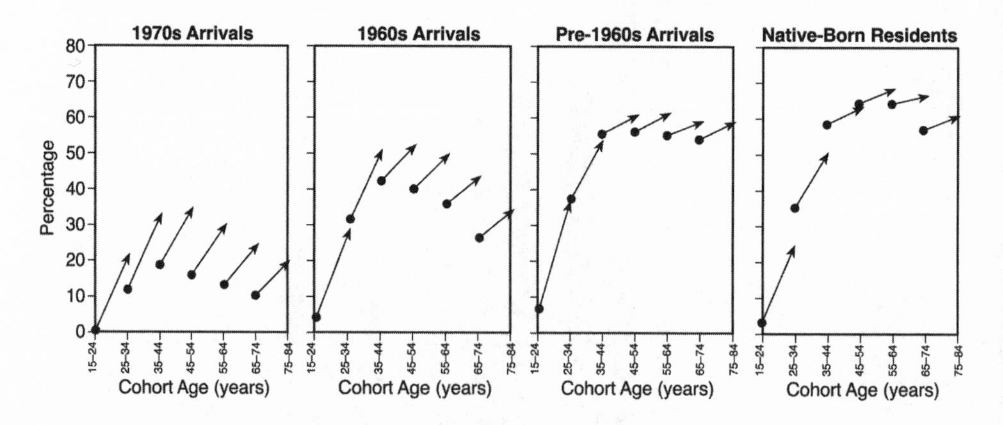

FIGURE 5.4. Double cohort: age and time of arrival in the United States for Mexican-heritage males who are single-family homeowners, 1980–1990. *Source:* Redrawn with permission from Myers et al., 1998.

results are an additional important finding about housing consumption of the foreign born. Not only are they able to penetrate the homeowner market, they are able to move upward within the ownership market.

OUTCOMES: HOMEOWNERSHIP RATES AND PATTERNS

The passage of time and the accumulation of assets creates homeownership—at different rates and to different extents across the United States. As the graphs suggest, ownership rates have risen for both the native born and immigrants. While immigrant groups have still not reached parity with the U.S. population as a whole, some foreign-born groups have averaged 6–7% increases in ownership levels (Table 5.2). Although some foreign-born Hispanic households lag behind other immigrant groups and their gains have not been as great, it is important to note that they are making gains and in many cases are outperforming native-born African American households as a group.

The contrasts in homeownership gains among specific groups are large and reveal different paths to ownership, reflecting demographic and contextual housing market effects. But the desire for ownership crosses ethnicity and background. The Aguilars bought their "dream home in Mariposa Isles (Florida), a new Miramar development targeting move-up buyers with floor plans that include mother-in-law quarters" (Elliott and Grotto, 2001, P. 1A). For the more than 120,000 Irvine (California) Asian residents the attraction is suburban safety, good schools, and high-tech jobs. It is not just the house but the "quiet cul-de-sacs and manicured lawns" (Cleeland, 1998, p. 1), or the neighborhood amenities that accompany high-end suburban developments, which are increasingly sought after by middle-class foreign born and native born alike.

Some states, notably Texas and Florida, have higher gains in immigrant ownership than do others, and within the states the ownership rates of particular groups are also different (Table 5.2). In California, the gains for the foreign born as a whole are quite modest, but this average gain masks quite different outcomes for Asians, who gain nearly 2% in levels of ownership, and for foreign-born whites, who actually lose ground. In New York, Asian foreign-born households make gains, but others have lower rates of ownership over time. White foreign-born households in Texas are now at the national average for ownership in the United States and have made real gains in the past two decades (Table 5.2).

TABLE 5.2. Homeownership Rates for Foreign Born by Year for the United States and the High-Immigrant-Impact Regions

	1980	2000
United States		
All foreign born	52.9	49.1
Asian	47.6	50.1
Hispanic	38.5	40.6
White	62.4	62.9
California		
All foreign born	48.1	45.8
Asian	51.7	53.6
Hispanic	35.9	37.3
White	60.1	56.7
New York/New Jersey		
All foreign born	43.0	36.4
Asian	34.5	45.2
Hispanic	20.8	19.5
White	52.4	51.9
Texas/Arizona/New Mexico		
All foreign born	53.2	53.9
Asian	50.3	43.8
Hispanic	51.2	53.3
White	61.7	67.5
Florida		
All foreign born	60.5	58.3
Asian	54.0	55.5
Hispanic	49.3	54.5
White	73.2	75.0

Source: U.S. Bureau of the Census, Public Use Microdata Sample, 1980, and Current Population Survey, 2000.

In Florida, foreign-born whites are also doing very well. However, it is true that some groups are only holding their own, and some foreign-born groups have declined slightly in their ownership levels—for example, white foreign-born households in New York. The data for Asians in Texas is based on a very small sample and may not reflect much change. Overall, then, the temporal patterns suggest that even in the context of expensive housing markets, immigrant groups—as is true for the U.S. population as a whole—are sharing in the oft-stated American goal of "owning your own home." It is true that the gains are not always large, but the evidence seems to be that ownership rates are increasing. Immigrants are doing what they report in survey responses—buying houses. Fannie Mae

(1995) reported that the highest priority for 61% of immigrants who rent is buying a house.

Currently, about 66% of all U.S. households own their home. About 48% of all foreign-born households are owners (Figure 5.5). This bracket defines a range within which we can examine the ownership levels of foreign-born ethnic groups for the United States and for the major immigrant regions in our study. For the country as a whole, white and Asian immigrants are above the national average and Hispanic and other origin immigrants are lower than the foreign-born average. These national patterns mask a great deal of variation across states. Only whites, and then barely, exceed the national foreign-born homeownership levels in New York/New Jersey, while white, Hispanic, and Asian groups all exceed the national average in Texas/Arizona/New Mexico and in Florida. In the latter, white foreign-born groups are significantly above the U.S. average as a whole.

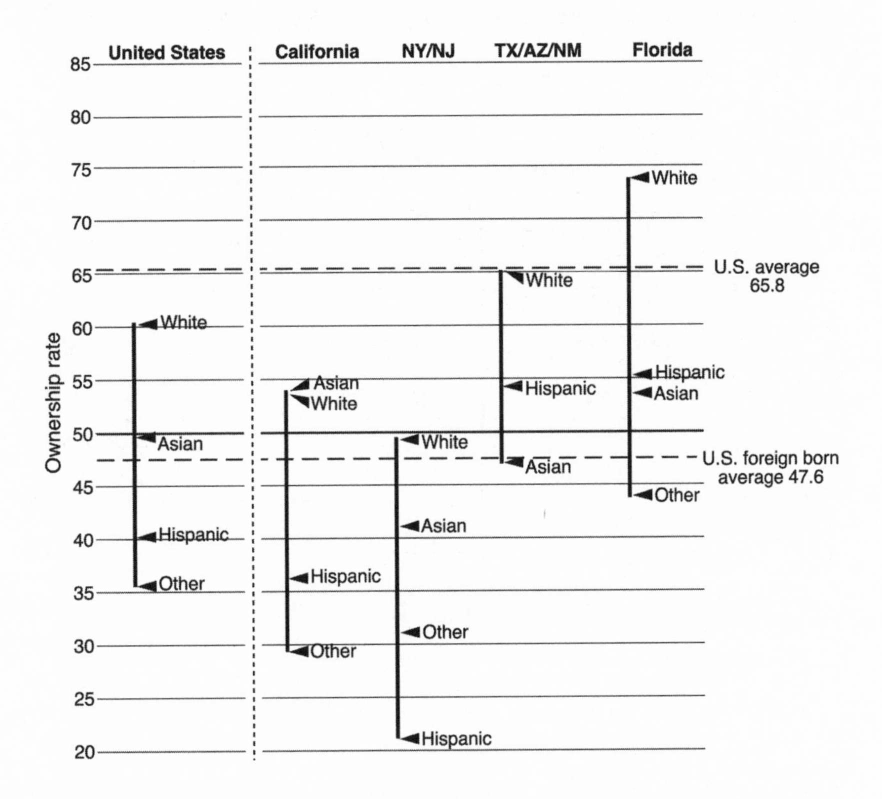

FIGURE 5.5. Homeownership rates for the foreign born in the United States and the four major-immigrant-impact states/regions. *Source:* U.S. Bureau of the Census, Current Population Survey, Combined File, 1998–2000.

Hispanic households in Texas and Florida are about halfway between average foreign-born rates and national rates, while Hispanics in California and New York/New Jersey have ownership rates which are some 20–30 points lower. In California, the combination of younger households and expensive housing markets has a powerful role in the levels of ownership (as this chapter explores later). In New York, the relative poverty of Dominican immigrants and the expensive housing market both play a role. It is clear that place matters, and to some extent so does ethnicity. White and Asian immigrant households do better than the Hispanic foreign born and immigrants from African and Caribbean origins.

When we control for period of entry, something that I showed was important in the foregoing discussion of trajectories, new stories about homeownership gains are revealed (Figure 5.6). For arrivals before 1980,

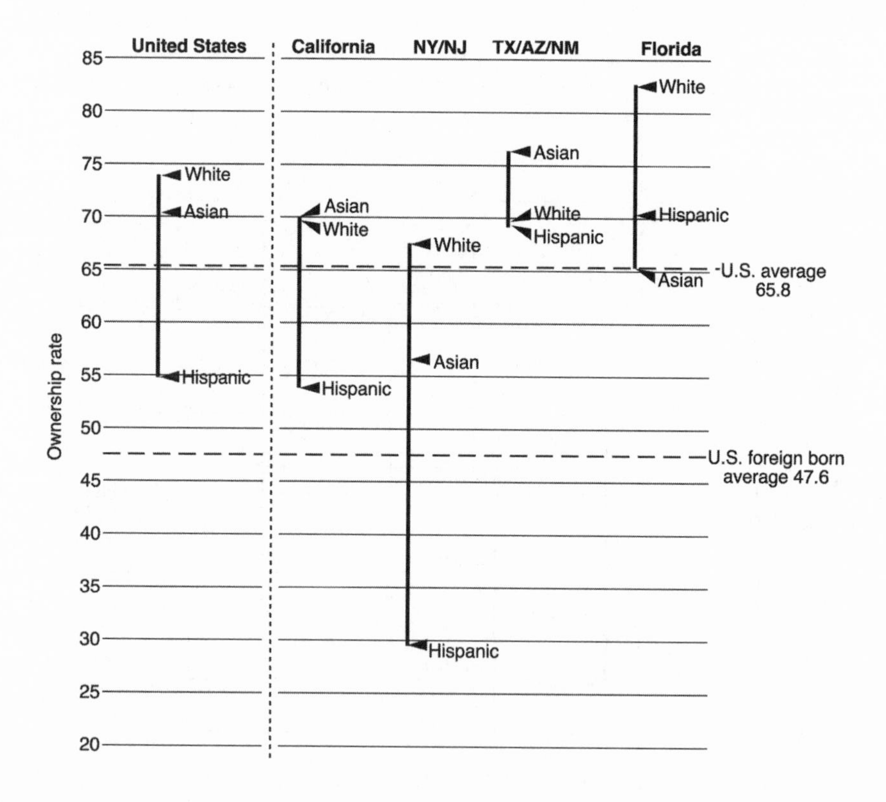

FIGURE 5.6. Homeownership rates for households who arrived in the United States by 1980. *Source:* U.S. Bureau of the Census, Current Population Survey, Combined File, 1998–2000.

almost all immigrants groups are above the national foreign-born average and many are above the national U.S. level. The shift in the ownership gains between Figures 5.5 and 5.6 is a testament to the ability of immigrants to gain a foothold in the housing market. While it is true that the ownership rates for Hispanics are lower in general, and lower in California, they are still well above 50%—half the households who arrived before 1980 are owners in 2000, some two decades later. Only in New York are Hispanic immigrant households doing poorly in the homeowner market.

Age effects were clear in the trajectory presentations, and they show how homeownership for foreign-born households increases steadily with increasing age (Figure 5.7). By age 50 the rates of ownership are at national average rates. Predictably, the ownership for Hispanics is lower and takes somewhat longer to reach nearly national average rates, whereas white foreign-born households are at national rates by age 40. Asian households have a slight decline in ownership rates at older age groups that may reflect cultural differences, or possibly the lower income

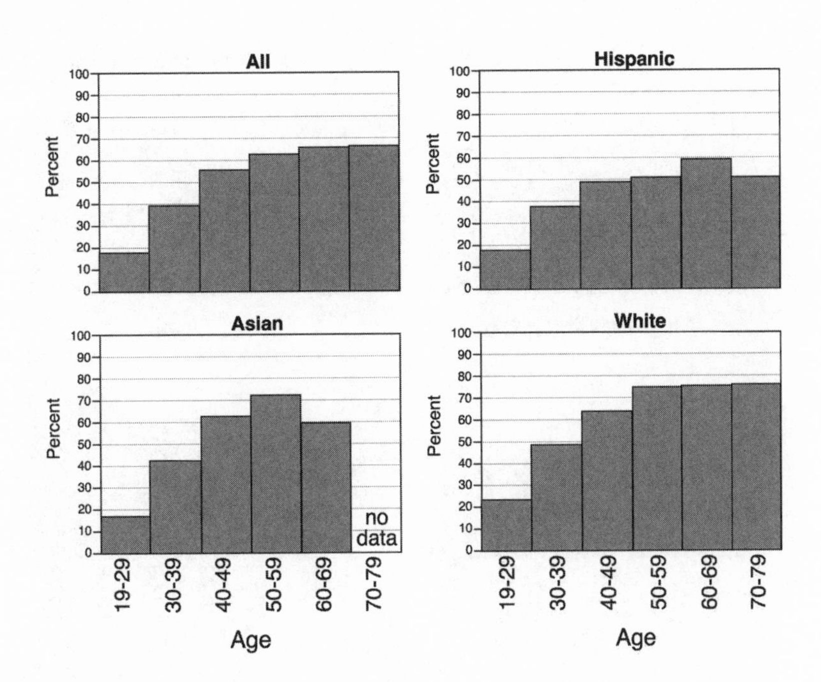

FIGURE 5.7. U.S. homeownership rates by age and ethnic background for foreign-born households. *Source:* U.S. Bureau of the Census, Current Population Survey, Combined File, 1998–2000.

levels of earlier refugee arrivals. The same patterns of increasing ownership with age are true for all the regions (Figure 5.8). However, the differences between the very high levels of ownership in Texas/Arizona/New Mexico and the lower levels in New York/New Jersey are striking. California and Florida are similar, although immigrants in Florida achieve somewhat higher rates overall. These graphs reiterate the importance of keeping the life course in mind. Average homeownership rates disguise what happens as immigrants stay and put down roots. Because so much recent immigration is of younger cohorts, average homeownership is necessarily lower and only by examining the age cohorts do we get a good picture of the ownership levels and the likelihood of immigrant households becoming owners.

Where immigrants choose to live affects their ownership rates. These contextual effects, particularly where people choose to live, show up in the different homeownership rates for Hispanics (Figure 5.9). Hispanic households by age 40 are pushing average national ownership rates in

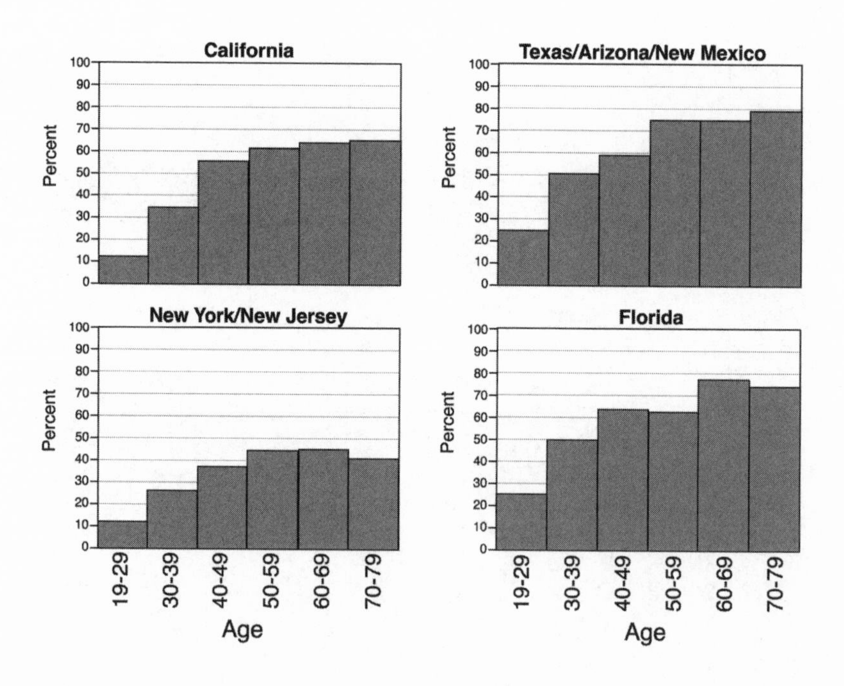

FIGURE 5.8. Homeownership rates by age for foreign-born households for major-immigrant-impact states/regions. *Source:* U.S. Bureau of the Census, Current Population Survey, Combined File, 1998–2000.

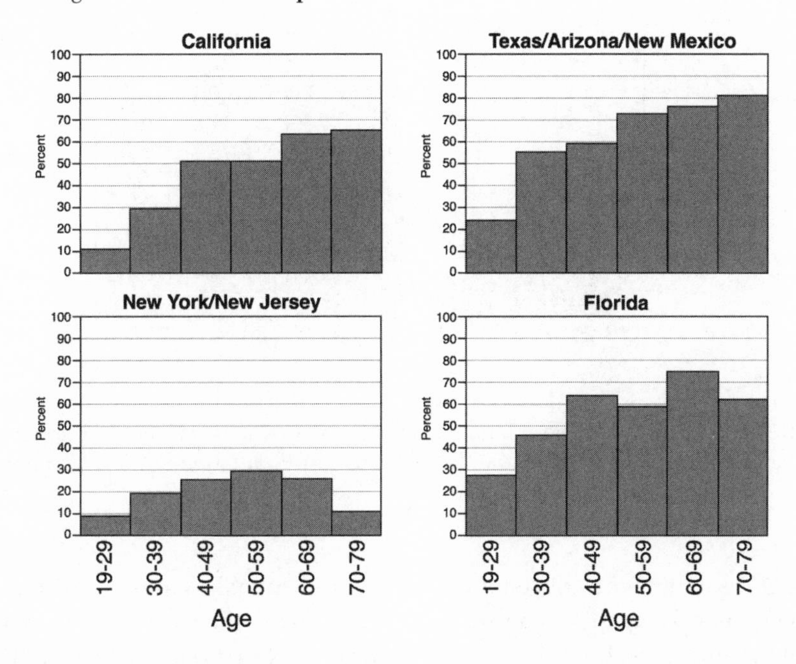

FIGURE 5.9. Homeownership rates by age for Hispanic foreign-born households. *Source:* U.S. Bureau of the Census, Current Population Survey, Combined File, 1998–2000.

Texas and Florida, and are above 50% in California. It is only in New York/New Jersey that ownership neither is high nor does it change a great deal with age; in fact, there the ownership rates show substantial declines after age 60. The Hispanic ownership rates in Texas/Arizona/New Mexico are quite remarkable, higher than in the other three graphs in Figure 5.9. The successful penetration of the homeowner market is a response to somewhat lower real estate prices there and relatively affluent Hispanic households. Average income for foreign-born Hispanic households in Texas/Arizona/New Mexico is nearly at the U.S. national average.

As much of the discussion has been focused on how middle-income immigrant households are doing, a specific graph portrays ownership rates by income and age (Figure 5.10). All income groups make gains in the homeowner market with increasing age, but the gains for middle- and upper-income households are greater and more rapid.[11] By age 30, both middle- and upper-income households are approaching national levels of ownership, and by age 40 both groups are at, or exceed, the national lev-

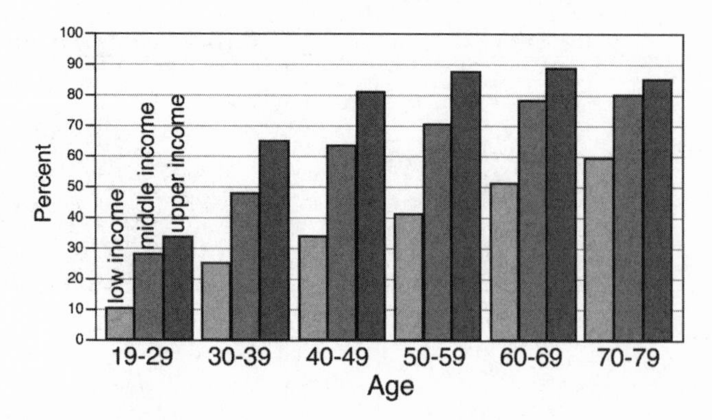

FIGURE 5.10. Foreign-born homeownership rates by low, middle, and upper income. *Source:* U.S. Bureau of the Census, Current Population Survey, Combined File, 1998–2000..

els. It is low-income households who are slow to make gains in the homeowner market, again not a revelation. Thus, given the recency of so many immigrant households and their youth, it is not surprising to find homeownership rates for low-income groups to lag significantly behind those of their wealthier compatriots.

There are few differences across ethnic middle-income households in the rates of homeownership (Figure 5.11). There is almost no difference between Hispanic and white middle-income households, although white households have slightly higher overall rates of homeownership. Inter-

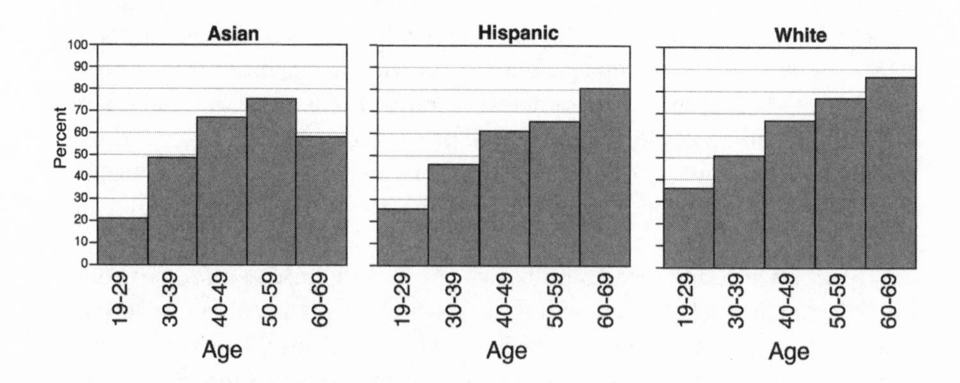

FIGURE 5.11. U.S. homeownership rates for the middle-income foreign born by age and ethnic origin. *Source:* U.S. Bureau of the Census, Current Population Survey, Combined File, 1998–2000.

estingly, older Asian middle-income households are less likely to be homeowners even though they clearly have the incomes that can support ownership. In this case the previous suggestion—that this is related to earlier refugee arrivals with lower incomes—is less relevant, and the results may reflect cultural differences and a lower desire for ownership. They may also represent a greater willingness of elderly Asians to live in apartments near to their homeowning children.

HOUSING MARKET EFFECTS ON HOMEOWNERSHIP

In general we expect new immigrants to enter the housing market on the lowest rungs of the housing ladder, and cross-sectional analyses tend to show new immigrants in the poorest housing conditions. Because the ability to satisfy housing needs is closely bound up with resources, immigrants who have just arrived and who may be less than fluent in English are more likely to be renters or to own lower-cost housing, often in the center of metropolitan areas. In addition, some immigrants who are members of ethnic minorities may encounter barriers to *buying* a home or obtaining high-quality housing with enough space, because of a lack of experience in home purchasing or because of discrimination in the housing market (Alba and Logan, 1992, 1993).

At local scales, complexity arises because the decision to own a home and the ability to make the transition from renting to owning are both dependent on local housing market conditions and in turn impact those same markets. On the one hand, local housing markets can benefit from the growth created by increased housing demand from immigrants; but on the other hand, there are increased costs of local growth—congestion, crowding, and price inflation.

Research by Alba, Logan, and Stutz (2000) is especially relevant here. They show that as their socioeconomic positions climb, members of all groups live in neighborhoods with more white and fewer minority residents, indicative of greater assimilation.[12] A part of this process of assimilation is closely tied to the ability to relocate within metropolitan areas. Language, citizenship, and economic status are all tied to the likelihood of making the upward and "outward" moves to suburban locations (Clark, 1998; Alba et al., 2000).[13] Evidence which supports these findings is found in a study by Galster, Metzger, and Waite (1999), whose results question the advantage of ethnic enclaves. Recall that an ethnic enclave occurs when there is a significant concentration of an (immigrant) ethnic group.[14] In Galster and colleagues' research, households in ethnic enclaves are negatively impacted by association with other coethnics—employment was lower and poverty higher. By extension, assimilation works: there are positive gains from being in neighborhoods and communities that are *not* dominated by coethnics.

Neighborhood and Community Outcomes

Immigrant progress in the housing market is clearly context dependent; that is, the ability of immigrants to move up in the housing market is dependent on the availability and costs of housing in the local markets where they arrive. Because much international migration is predominantly an urban phenomenon, most immigrants settle initially in large cities and only slowly move out from these entry-port environments. Moreover, these entry ports are often "global cities" (see below), interconnected through a vast web with other such global metropolises.

While New York and Chicago still receive large numbers of immigrants, the new immigrant centers are the four main gateway cities of the Pacific, Los Angeles, San Francisco, San Diego, and Seattle; the Caribbean metropolis, Miami; and the inland metropolitan area, Dallas–Fort Worth. These cities are the enters of new immigration, and it is in them that the housing markets are being transformed by the sheer numbers of new foreign-born residents.

The globalization that followed the deregulation of Western economies in the 1980s and 1990s has magnified transborder economic relations, much of it through the large metropolises that are the national headquarters cities (Ley and Tutchener, 2001). As the discussion in Chapter 1 showed, beginning in the 1970s and accelerating in the 1980s, the world economy has been marked by increasing interconnection among the very largest cities. These "global cities," with close ties to international flows of capital, are now global markets for offices, hotels, and condominiums, and by extension for real estate generally. The property deals in the global cities have had, in turn, spillover effects on the residential markets of these large metropolitan areas. More permeable borders and the internationalization of the world economy have created a situation in which price movements of housing show both geographic and historical synchronicity with globalizing trends. A case study of Canadian cities shows that the price takeoff in the housing market is closely linked to the rapid increase in immigration numbers in the mid 1980s (Ley and Tutchener, 2001, p. 220). Similar findings are true for large cities in the United States as here too the growth in net population is largely immigration driven, and because immigration is such a large proportion of all population growth, it necessarily must be tied to housing price changes.

California real estate markets boomed in the 1990s in conjunction with the very rapid increase in immigration. Median house prices rose dramatically in just a few years. In some cases in San Francisco, the most volatile of the local real estate markets, the prices doubled in a 2-year period from 1998 to 2000 (Federal Reserve Bank of San Francisco, 2000). Much of the increase in the San Francisco region is in the Silicon Valley area, where the rapid development of the high-tech sector has led to increased demand for housing. What was once a relatively affordable housing market in San Jose, where immigrants could buy less expensive housing, is now considerably more expensive. Increases in stock-market wealth have been translated into increasing house prices. To reiterate an earlier discussion, there seems to be a strong case for the association between the shifts of global capital and increasing local housing market costs in the new global cities. The patterns in San Francisco and to a lesser extent in Los Angeles can be seen as outcomes of an increasingly linked world. It is also true that increasing numbers of immigrants themselves increase competition for housing, and so drive up house prices.

The finding of a positive correlation between average house prices and immigrant proportions in both Canadian (Bourne, 1998) and Australian urban places (Burnley, Murphy, and Fagan, 1997) can be interpreted as a pressure on housing and by extension on prices. Affordability problems seem to have worsened, and a continuation of the present situation would put homeownership out of the reach of nearly all migrants in these large urban centers (Ley and Tutchener, 2001). It is not surprising that the proportion of foreign born who are renting has increased significantly in the past decade. Refugees and other low-income immigrants are particularly disadvantaged in very expensive housing markets like those in Vancouver, British Columbia, Toronto, Ontario, and U.S. gateway cities.

In addition to problems of affordability, immigrant families in general are larger, both because of higher fertility rates and more extended family structures, than their native-born counterparts. Larger family sizes further complicate the process of acquiring satisfactory housing space. An extension of not being able to afford homeownership is that immigrants cannot live in the areas with adequate social and community services.[15] Thus, the influx of immigrants has increased the demand for housing across particular metropolitan communities of the United States and reduces the housing opportunities available to them. Recall that nearly 20% of the total household growth in the United States was due to immigration. In addition, this increase is concentrated in a dozen metropolitan areas in half a dozen states—especially in New York/New Jersey and in California.

Two recurring themes emerge in discussions of the nature of immigrant housing: overcrowding and central concentrations. Immigrants, particularly renters, more often occupy overcrowded housing than do the native born (Myers, Baer, and Choi, 1996).[16] Yet crowding in rental accommodation is, for some, a deliberate strategy to pool resources for later access to the homeowner housing market or to remit funds to prospective immigrant family members in the country of origin. Indeed, immigrant families from certain cultures are inclined to live with extended family members and care for elderly relatives (Pader, 1994; Gove, Hughes, and Galle, 1983).

The proportion of households with a room shortage, defined as the difference between the actual number of rooms available and the required number of rooms, which is based on a formula related to family size,[17] is higher in counties with large numbers of immigrants (Clark, Deurloo, and Dieleman, 2000). Low-income households in high-immigrant-impact counties are almost four percent more likely to experience a room shortage than are the same-income-level households in low-immigrant-impact counties. It appears that there is greater competition for housing space in counties with large numbers of new low-income immigrant households. That said, it is clear that it this an income-based finding. The three lowest-income deciles are more than twice as likely to experience inadequate housing space than are middle- and higher-income households (Clark et al., 2000). Still, of the lowest-income households, *nearly half are able to move out of crowded housing when they relocate.* The finding that even the lowest-income crowded households can move out of crowded conditions is a finding of the gains that are possible in the U.S. housing market and a further affirmation that, even within the pressures on housing space, immigrants are still making gains.

It is clear that there are multiple paths and varying contexts for entry into the housing market, it is the intersection of those contexts and paths which explains the levels of homeownership attainment by differing groups of immigrants. An illustration from California's gateway cities will heighten our understanding of the complexities.

Gateway Cities in California

Gateway cities in California are among the most ethnically diverse cities in the world. Los Angeles no longer has a majority of any ethnic group, and San Francisco is rapidly changing its ethnic composition. The changes are a direct result of the large-scale immigration into California in the past three decades. The metropolitan populations are a little more than 12 million in the Greater Los Angeles region, and 5.7 million in the metropolitan San Francisco area. The foreign-born population is 37% in Los Angeles and 26% in San Francisco. Hispanics are the largest group in Los Angeles, and a mix of immigrants from Asian nations dominates the foreign-born population in San Francisco.

Following the earlier discussion of affordability, it is worthwhile examining how immigrants are doing in these large gateway cities. Clearly, the housing markets in the gateway cities are difficult for new arrivals, immigrant or not. The median housing values are significantly higher than for the whole United States (Figure 5.12), and about 40% of the housing market pays more than 30% of family income for housing. There is a significant rent burden/housing cost in these two cities for the population as a whole, and given the lower incomes of immigrants, this is even more true for them (Clark, 1998). In addition, house prices have been increasing in both cities, although most dramatically in the San Francisco Bay area. The increases in San Francisco have made ownership difficult for the population as a whole.

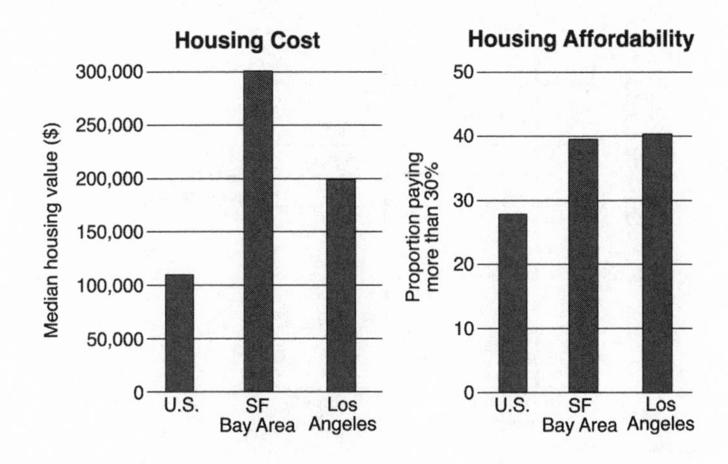

FIGURE 5.12. Median housing values and affordability in two gateway cities in California. *Source: American Housing Survey*, 1999.

Just how difficult is it to be a homeowner in these gateway cities and, by implication, in other high-immigrant-impact cities? Despite rising prices in these cities, can immigrants still make gains in homeownership? And can the foreign born achieve homeownership at native-born levels of ownership? Recall that the average ownership rate is about 66% for the United States as a whole. The foreign-born ownership rate is nearly 20 points lower, at 48%. No foreign-born group in Los Angeles or San Francisco reaches national levels of ownership, though foreign-born homeownership rates are higher in San Francisco than in Los Angeles (Figure 5.13). Clearly, housing costs *alone* are not determining the ownership rates of the foreign born. Asian and white foreign-born ownership rates are significantly above the rates for Hispanics in both cities. Asian and white ownership rates are edging toward the average rates for the United States in housing markets that are much more costly than those in most other metropolitan areas. Hispanic households lag most obviously. They own at rates of about 40% in San Francisco and just above 35% in Los Angeles. However, recall that this is a proportion of all households, including the most recently arrived. An analysis of deviations (from the U.S. average for the foreign born) for specific subgroups in California's

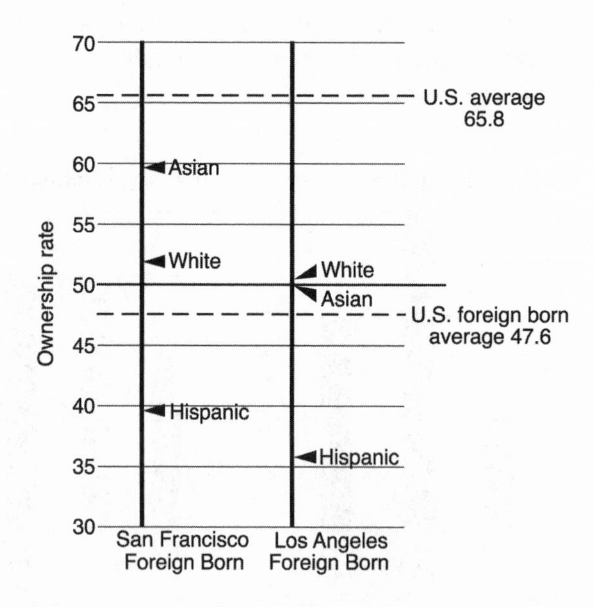

FIGURE 5.13. Homeownership rates in two gateway cities in California. *Source:* U.S. Bureau of the Census, Current Population Survey, Combined File, 1998–2000.

two large gateway cities reveals a broad spectrum of experiences among the foreign born.

While it is not possible to calculate homeownership rates for all groups because of small sample sizes, it is clear that Asian households are doing well in San Francisco. For those groups for which we have data, all are at or above U.S. average rates. Unusually, because in general Central American households are struggling, in San Francisco they are well above the U.S. average rates (Figure 5.14). Los Angeles has some Asian households with significant deviations above the U.S. average, notably the East Asian and Korean foreign born, but the rest are below the average rates. These results suggest some caution in arguing for successful homeownership and assimilation across all groups. However, the issue is really one of how the groups are doing *over time* and, even more importantly, how older immigrants are doing and what are the ownership rates for those

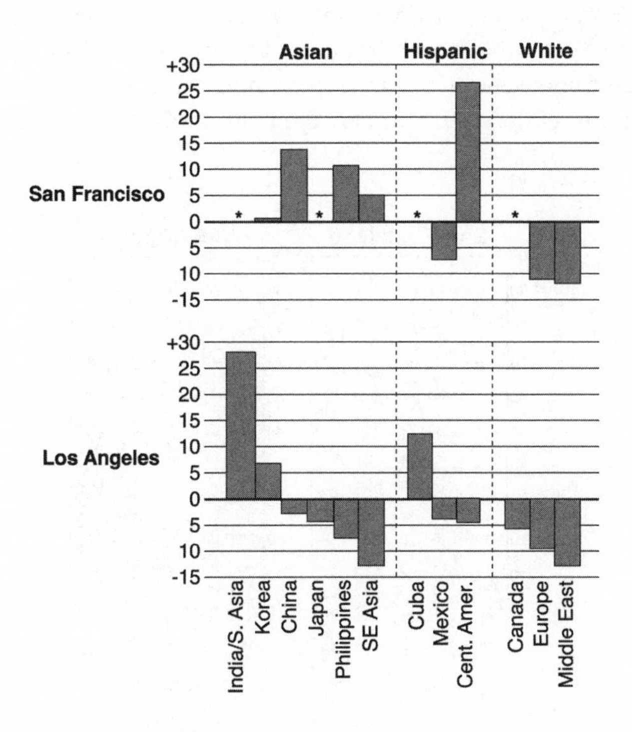

FIGURE 5.14. Deviations in homeownership rates for selected ethnic origins in two gateway cities in California: percent deviations from U.S. foreign-born averages. (*Data are not available for all groups.) *Source:* U.S. Bureau of the Census, Current Population Survey, Combined File, 1998–2000.

who arrived a decade or two ago. To reiterate, many recently arrived young immigrants will naturally depress the ownership rates.

As for the population in general, immigrant homeownership increases with age in both San Francisco and Los Angeles (Figure 5.15). The foreign born are at or above 50% ownership rates by the time they are in the 40–50 age group, and by the next age cohort they are nearly at the national average in San Francisco. The increase with age is a striking confirmation of the same housing progress that is true for the native born. It is true that the increase is slower in Los Angeles and the same levels are not achieved. Even so, that immigrants are making gains with age in both cities in the same way as the native born do and that older age groups in San Francisco are approaching average national levels is a testament to the general process of ownership gains over time.

A detailed analysis of ethnic groups in Los Angeles, the gateway city where the sample size is large enough for a more detailed analysis, reveals the same pattern of increased ownership with age. As expected, Asian and white foreign-born immigrant households follow a path of relatively rapid increases in ownership with age (Figure 5.16). Asians very quickly get to national levels, and their rates only fall off for older age

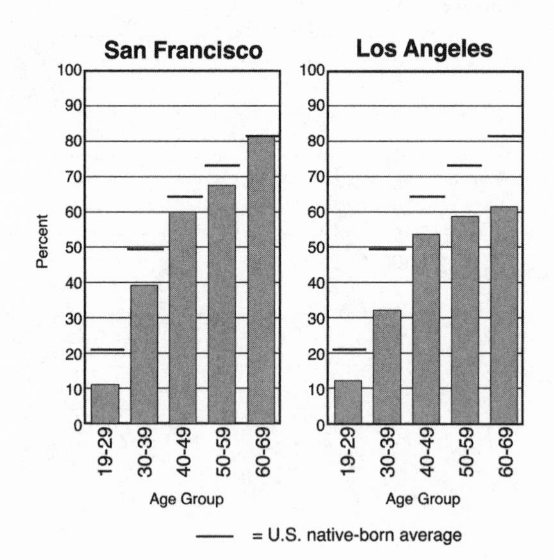

FIGURE 5.15. Homeownership in San Francisco and Los Angeles for the foreign born by age groups. *Source:* U.S. Bureau of the Census, Current Population Survey, Combined File, 1998–2000.

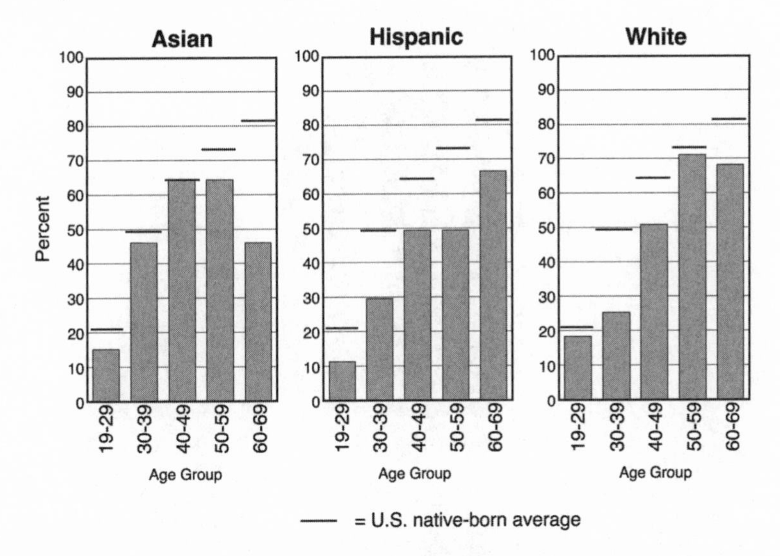

FIGURE 5.16. Homeownership by ethnic origin in Los Angeles by age groups. *Source:* U.S. Bureau of the Census, Current Population Survey, Combined File, 1998–2000.

groups. Hispanics gain in the middle-age cohorts, and whites start out slowly but make rapid gains. It is true that Hispanics lag in Los Angeles, but the truly significant finding is that the gains in ownership extend through late adulthood into the 60–69 age cohort, when they are around the national average for the United States as a whole. Clearly the process of ownership increases with age is working, even if more slowly, for the least-advantaged groups.

An additional finding which provides further support for an optimistic interpretation of the ownership process is the data on ownership for immigrants who entered in earlier decades. Immigrants who arrived before 1980 have ownership rates that are around, or above, the national native-born averages. Those rates decrease for later arrivals. This finding is true across immigrant groups in San Francisco and Los Angeles, but again Hispanic immigrants have slightly lower rates of ownership (Figure 5.17). Yet, those rates are much closer than we have seen when we examine the ownership rates by age. While the ownership rates decline over time for all groups and are significantly lower for recent arrivals, it is the big decline for Hispanic households who arrived after 1980 which is most notable. Overall, the graphs and tables reveal the very robust nature of the owning process. Earlier arrivals and older immigrants—in essence

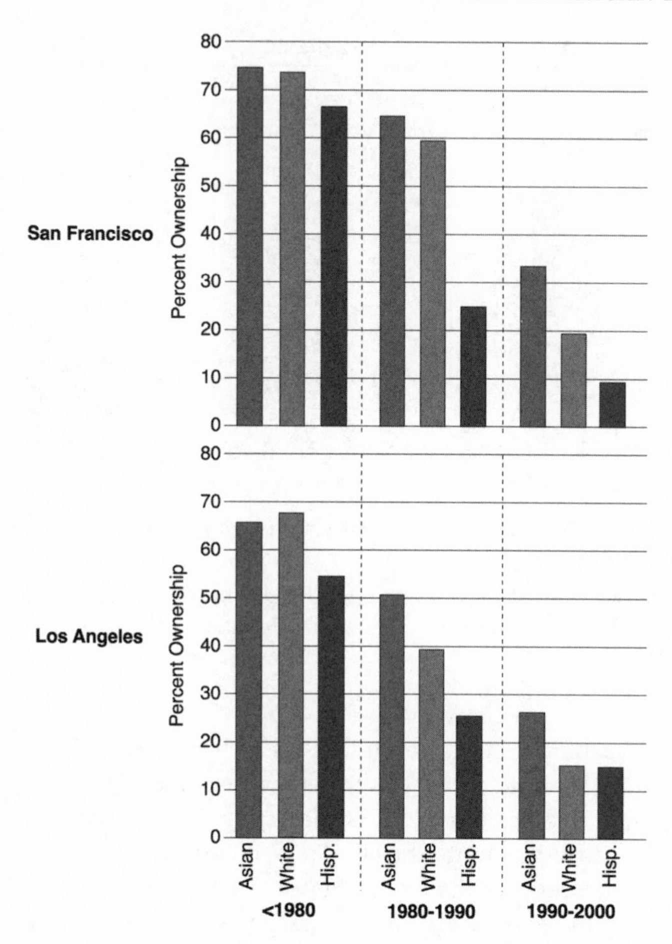

FIGURE 5.17. Homeownership rates by period of entry into the United States for Los Angeles and San Francisco. *Source:* U.S. Bureau of the Census, Current Population Survey, Combined File, 1998–2000.

those who likely arrived earlier—are becoming owners at a substantial rate.

All in all, social integration as measured by ownership is substantial and widespread across different groups and different metropolitan areas. To reiterate the earlier discussion of assimilation, buying a house is putting down economic and social roots. It is an investment in the new society, and by selecting particular neighborhoods, it is a part of the process of integration. While it is true that many first-time homebuyers are purchasing houses in inner-city neighborhoods, among homebuyers from similar ethnic backgrounds the preliminary data on spatial patterns of home purchase in Los Angeles reveal a pattern of large-scale suburbanization of home purchasing too (Clark, 1998). The other important finding is that the homebuying process is occurring in what are very expensive homeowner markets. The high prices in these markets have not deterred foreign-born households from advancing within them just as earlier waves of immigrants did.

Is the story all positive for home purchasing? Clearly more recent and younger immigrants are less likely to be owners and may well have a more difficult time in making the progress that earlier arrivals have achieved. Homeownership for poor households (defined as households by income below the official poverty line) in Los Angeles was just 26.7% (*American Housing Survey*, 1999). In addition, there is a shortfall of approximately 400,000 affordable rental housing units (American Housing Survey, 1999). The Southern California Association of Governments (SCAG) estimates that nearly 200,000 new housing units will need to be built in Los Angeles County by 2005. These findings suggest the difficulty faced by immigrant households trying to establish a foothold in the housing markets of gateway cities in California, and by extension in the United States as a whole. At the same time, that so many households have become owners is a testimony to the power of the dream of ownership and the tenacity of new immigrants in these gateway cities.

SUBURBANIZATION AND SPATIAL ASSIMILATION

Much of the interest in homeownership stems from an interest in how the new immigrants are assimilating. Are they buying houses and thereby integrating into communities? Apparently so—as increasing ownership can be used as an indication that immigrants are "putting down roots." A vigorous debate centers on the issue of whether new immigrants, especially those who arrive with fewer skills and little or no capital, will follow the paths of earlier immigrant groups toward assimilation and integration or whether they will form a segregated underclass community in the central cities of American metropolitan areas. Thus a crucial part of this assimilation process relates to *where* immigrants are purchasing houses and putting down roots as homeowners.

Overall ownership rates for the foreign born are much higher in the suburbs than in the central city (Table 5.3). The ownership rates for all for-

TABLE 5.3. Foreign-Born Homeownership in Central Cities and Suburbs in the United States

	Central cities	Suburbs
All foreign born	37.3	56.8
Asian	45.0	62.1
Hispanic	31.1	46.2
White	43.2	70.0
Other	33.0	46.9

Source: U.S. Bureau of the Census, Current Population Survey, Combined File, 1998–2000.

eign born who live in the suburbs are about twice that of the foreign born who live in the central city of large metropolitan areas. It is what we would expect as new arrivals head for the city centers where rental housing abounds and where there are large numbers of coethnics making their way in the new society. Similar patterns exist for the different ethnic groups as well. Additional support for the notion of integration and ownership is supplied in the much higher ownership rates for the foreign born who move from the city to the suburbs. The ownership rates for these transfers are three times as great as for those who move within the inner city (Table 5.4). Among mobile foreign born from all ethnic divisions, those relocating from the city to the suburbs are significantly more likely to be homeowners compared to those changing residence within the same city. Clearly, the rates of ownership are lower in the city because it is there that the new immigrants are congregating. These immigrants

TABLE 5.4. Homeownership Rates by Residential Moves in U.S. Metropolitan Areas

	Within CC	CC to suburbs	CC to nonmetropolitan areas
All foreign born	22.8	41.8	56.2
Asian	25.7	49.2	—[a]
Hispanic	20.3	37.2	49.0
White	24.3	49.2	48.4

Note. CC, central cities. *Source*: U.S. Bureau of the Census, Current Population Survey, Combined File, 1998–2000

[a]Small sample size.

have fewer assets and lower levels of human capital, which in turn lead to lower incomes. But just as surely, those who can accumulate assets and increase their income are moving to the suburbs and becoming homeowners.

At the same time, there is evidence that the suburbanization–assimilation link is not always straightforward. A national study of homeownership and central city and suburban locations concluded that suburbanization may be related to factors other than simply socioeconomic resources. Fong and Shibuya (2000) found that immigrant households that were more acculturated and that have more socioeconomic resources are more likely to be homeowners in the central city than to live in the suburbs as renters. Even so, this finding is not necessarily counter to the findings in California and the whole United States. Ownership is likely to be, on average, less expensive in the central cities, and immigrants may well make their initial move into ownership in the central cities and only later follow this with a move to ownership in the suburbs. Preliminary data from the 2000 U.S. Census for the Los Angeles metropolitan region suggests that ownership is increasing among Hispanics.[18] The data also suggest that Latino immigrants in Southern California first rent in the urban core, and then after one or two decades they disperse to the suburbs, where they buy homes and "assimilate" into those suburban communities. Both San Bernardino and Riverside counties have had very large gains in Latino ownership in the last decade. These gains are exactly what we would expect of an assimilating population, seeking out affordable suburban ownership in the growing fringes of the metropolitan area. The data from the 2000 Census shows that Latino ownership in San Bernardino and Riverside counties increased by 55 and 50%, respectively.

Two contrasting stories of Latino ownership add to the insights gained from the detailed analysis of the Census data. The Quintes family was attracted to San Bernardino County because they could buy a three-bedroom home for $135,000 with only 3% down. At first they commuted to their small business in Los Angeles, but within a year they moved it to Rancho Cucamonga, only a 15-minute drive from their home in Fontana. Houses are more affordable in Fontana than in Los Angeles, they said, and "the quality of life is better" (*Los Angeles Times,* August 27, 2001, pp. B1–B2). In another instance the Martinez couple bought a four-bedroom home in Highland Park for $195,000. They chose to pay more to be closer to their extended family and to their jobs. Such stories provide us with useful vignettes that reveal the lives behind the statistics: although Highland Park is hardly central-city Los Angeles, the decision of the Martinez couple to take the commute and their family's needs into account is a classic response of reducing the length of the journey to work by accepting the trade-off with more costly housing; in contrast, the Quintes family opted for more affordable housing in Fontana, and later relocated their business from Los Angeles to nearby Rancho Cucamonga.

WHO OWNS?: EXPLANATIONS OF SUCCESS IN THE HOMEOWNER MARKET

Accessing the housing market is a function, obviously, of a family's assets and its income. Higher and more stable incomes make homeownership more accessible, as do the collective assets that an extended immigrant family can pool. But the latter tends to be beyond measurement, so we must focus on the former. It is expected that those with higher incomes and those with more human capital (greater levels of education), which can be translated into professional occupations and in turn higher incomes, will be more likely to be owners. The model to investigate ownership by the foreign born uses these variables to explain ownership. In addition, I argue that U.S. citizenship, another measure of assimilation, is related to the likelihood of becoming an owner. In a sense the transition to U.S. citizenship is another measure of commitment to the new society. As the earlier descriptive analysis documented, age and time of arrival play an important role in becoming an owner, and they are also included in the model.

From the discussion of trajectories to ownership and of the key dimensions which make ownership possible—age, income, and time of arrival—we can formulate a simple model to assess how these variables interact and which are the most important. A simple log odds model, in which the ratio measures the probability of ownership as a function of the variables, is transparent and easy to understand. For example, an odds ratio of 2, for U.S. citizenship, would indicate that being a citizen would double the probability of being a homeowner.

The model for the United States and for California, New York/New Jersey, Texas/Arizona/New Mexico, and Florida is significant and has concordant ratios that range from 77 to 80%. That is, the model correctly predicts, on the basis of the independent variables, who is likely to be a homeowner (Table 5.5). The log odds ratios can be compared for the United States or across states and regions. They are measures of the effect of the independent variables. They provide a collective quantitative sum-

TABLE 5.5. Predicting Homeownership of the Foreign Born

	US	CA	NY/NJ	TX/AZ/ NM	FL
Age	1.03	1.03	1.02	1.03	1.03
Income	1.00	1.00	1.00	1.00	1.00
College education	1.26	1.32	1.61	0.81	1.54
Married	2.91	2.60	3.12	3.00	2.32
Citizen	1.65	2.01	1.59	1.52	1.66
Arrive < 80	3.46	3.77	4.72	2.72	2.71
Arrive 80–90	2.02	2.16	2.60	1.81	1.95
Percent concordant	79.8	81.3	81.3	77.5	77.3
Tau	.298	.310	.284	.274	.266

Note. Values in the table are odds ratios. Thus, a ratio of 2 indicates that, for example, being a U.S. citizen in California doubles the probability of being a homeowner.

mary of our empirical descriptive analysis of both age and time-of-arrival effects.

In every state/region, the time of arrival is one of the strongest predictors of homeownership. Immigrants who arrive before 1980 are three or more times likely to own their homes than are more recent arrivals. Even those who arrived in the 1980s are twice as likely to own their homes as are the most recent arrivals. There are some interesting variations across the states/regions, and other variables play an important role in the likelihood of ownership. In California, U.S. citizenship is more important than in the other states; in New York/New Jersey, college education is an important factor. In New York/New Jersey and Texas/Arizona/New Mexico, marital status is important and significantly raises the probability of ownership. The coefficients for income are significant, but note that the odds ratios are 1 when income is a continuous variable. The results reemphasize the impact of length of residence. By extension, the models provide a basis for arguing that it is time of arrival and assimilation that are critical in the homeownership process.

CONCLUSIONS AND OBSERVATIONS

The discussion and analysis in this chapter provides an update of earlier research on homeownership rates for the foreign born and enriches it with both national and local outcomes. The research goes beyond previous work, arguing that there is a strong link between homeownership and integration. The evidence is partly circumstantial; we cannot draw a direct link between ownership and integration, but the strong links between time spent in the United States and homeownership, between time of arrival and ownership, and between suburbanization and ownership are clearly suggestive of an ongoing process of assimilation. It is a process that involves increased community identity as immigrants become owners. In addition, the links between citizenship and education are indeed evidence of the way in which increasing participation (evidenced by U.S. citizenship), homeownership, and increased human capital are linked. Family formation (marital status), too, plays a key role in the ownership process.

It is not possible to overemphasize the evidence for an integrative process that comes from the data on time of arrival. Despite the concerns about new poor immigrants, those who arrived in earlier periods are doing what immigrants always did—integrating and becoming part of the mainstream and, when they can, buying a part of the American Dream. Whether the same process can work for new waves of immigrants will depend on both the changing economic contexts and the extent to which there is a commitment to affordable housing in U.S. metropolitan areas.

NOTES

1. Other studies of immigrant success have also stressed the important role of homeownership for new immigrants (Bourassa, 1994; Alba and Logan, 1992). Some have shown that movement into homeownership can be quite rapid (Clark, 1998; Myers and Lee, 1998).
2. The U.S. Department of Housing and Urban Development suggests that every new home creates 2.1 jobs directly and many more indirectly.
3. G. McCarthy et al. (2001) point out that the benefits of homeownership are not distributed evenly among homeowners. Lower-income households are more susceptible to investment losses in housing, and those homeowners who use "affordable" mortgage instruments are more highly leveraged than other homeowners. Perhaps most critically, lower-income households incur higher maintenance costs because they purchase lower-cost homes in the center of cities and thus are more likely to be living in older homes. There is evidence, too, that lower-income households tend to have higher transaction costs.
4. During the 1970s and early 1980s house prices in major U.S. metropolitan areas appreciated at about 8% a year—a significant return, although less than the Standard & Poor's (S&P) 500 Stock Index. During the later 1980s and the

1990s annual returns on housing in the metropolitan areas of Boston, Chicago, and Los Angeles were about 6.5% (G. McCarthy et al., 2001), and generally these returns are much less risky than stock market returns. There are risks in housing, as in any investment, and high leverage may decrease the liquidity of housing assets. Certainly, highly leveraged households will have less flexibility with respect to sales and refinancing (Archer, Ling, and McGill, 1996).

5. Overall, as compared to homeowners, renters are twice as likely to suffer from physical deficiencies and three times more likely to live in crowded conditions (G. McCarthy et al., 2001). Homeowners live in larger units, on average about a third larger than rental units, and are likely to have more than one bathroom, whereas three-quarters of rental units have only one bathroom or do not have access to a bathroom within the rental unit (G. McCarthy et al., 2001).
6. I recognize that using the language of adaptation will offend some multiculturalists: why should immigrants have to "fit in," they will ask, but I intend the argument to be about adaptation and the advantages that adaptation will bring, advantages that I believe are greater than the division into separate societies.
7. A similar analysis can be created when the 2000 PUMS data are released. It is likely that the patterns if not the exact levels of ownership will be re-created in the 2000 PUMS data as are shown here for the changes from 1980 to 1990.
8. When the 2000 detailed Census data are released, it will be possible to examine the changes through three decades.
9. It is possible, too, that the small numbers for these immigrant groups generate these results.
10. The data and graphs are drawn from presentations in Myers et al. (1998). The graphs have been redrawn and aggregated into a single figure.
11. The middle-income range is from $34,058 to $84,975 in 2000 dollars. Upper-middle and lower-middle ranges are divided at $59,517.
12. Although Alba et al. (2000) provide some support for the asssimilation argument, they also suggest from their longitudinal models that the power of the assimilation model is not as striking in the 1990s as in earlier decades. They attributed this to the very-large-scale immigration of lower socioeconomic immigrants in the last decade of the 20th century.
13. A specific study of suburbanization and homeownership by Fong and Shibuya (2000) does raise questions about the spatial assimilation perspective. Although they acknowledge that while households that are more acculturated and have more socioeconomic resources are more likely to be homeowners, they point out that ethnic groups are more likely to be homeowners in the central city than in the suburbs, casting doubt on the suburbanization spatial assimilation perspective. A simple counter to that argument, however, is to note that most of the less-expensive housing is in central cities and that is where we would expect to find first-time ethnic homebuyers.
14. There is no agreed definition for the proportion of an ethnic group necessary

to constitute an enclave, but many working definitions assume a proportion greater than 70%.

15. In a study of housing in Sydney, Australia, a city with a very high foreign-born population, Ley and Murphy (2001) show that recently arrived immigrants were significantly more likely to suffer housing stress (defined as those with incomes in the lowest 40% of the income distribution who are paying more than 25–30% of their income on housing) than were older immigrants or the native born.
16. Most crowding studies use a criteria of more than one person per room. A rooms stress or adequacy measure may be a better way of assessing crowding. I use that measure later in the discussion.
17. The required number of rooms defined in the Panel Study of Income Dynamics for the United States is generally two rooms for each household head (including a spouse) and additional rooms depending on the number, age, and sex of additional occupants (Clark et al., 2000).
18. See D. Wedner, Education, employment gains help more Latinos become homeowners. *Los Angeles Times,* August 27, 2001, p. B1).

CHAPTER 6

★ ★ ★

Voicing Allegiance

A sea of tiny American flags waving at a naturalization ceremony is the symbolic expression of the commitment and attachment of immigrants who have made the transition from "aliens"[1] to citizens. This attachment to the United States, by becoming a naturalized citizen, is an important measure of participation, and by extension of assimilation into U.S. society. Becoming a citizen is evidence of a new allegiance, being a part of America, even of "Americanization."[2] While sociologists may debate the terminology of Americanization, the notion resonates with many new immigrants who have come to make a new life in the United States. Either they or their children will undergo some form of incorporation or assimilation over their lifetime.

Clearly, incorporation is multidimensional: new immigrants assimilate across economic, social, cultural, and political dimensions and at varying speeds and intensities. And just what assimilation is varies from the perspectives of the native born and immigrants, from person to person, and from group to group. The previous chapters examined the occupational and housing dimensions of integration, of becoming more like the native born in professional status and in levels of homeownership. In this chapter I turn to the political dimension. After all, becoming a citizen and voting are central measures of just how and where the new immigrants are participating in U.S. society. To explore political incorporation, the chapter examines both who naturalizes and the interaction of naturalization and political behavior.

The central aim of the chapter is to show, consistent with the arguments on professional gains and entry to homeownership, that naturalization is occurring even though there are ethnic variations in the pace of naturalization. Certainly the data which follow cannot be used to argue that newcomers are doing something different from those in the past; moreover, the evidence suggests that the pace of political integration as measured by naturalization may be increasing.

While immigration can be a reactive response to the lack of economic opportunities in the home country or a response to political repression or persecution there, naturalization is a proactive response. It is a decision to become a part of a new society, a decision to participate fully in the collective civic processes of the society. The limited evidence from surveys of newly naturalized citizens suggests that, once naturalized, they are more likely to take up active roles in the civic life of the country, to vote, to join community action groups, and to be active in politics more generally (Bouvier, 1998; Portes and Rumbaut, 1996).

But again, as in so many of the subtexts about immigration and its outcomes, there are strongly held views on the political consequences of admitting large numbers of new citizens. On the one hand, those who favor large-scale admissions of new immigrants argue that these new residents indeed want to participate in their new society. The proponents of incorporation argue that, with specific policies to facilitate it, the input of these new residents will enrich the American democratic process (Plotke, 1997). Moreover, to the extent that citizenship is easier to attain at the beginning of the 21st century than it was a century ago because of expanded suffrage, civil rights legislation, and a society which is less exclusionary, we might expect even greater movement to incorporation and assimilation. On the other hand, there are equally strong arguments from those who suggest that the "nature and legitimacy of the nation is being challenged internally by multi-culturalism and alienation" and a movement away from a common national identity (Pickus, 1998, p. 109). Debates about the nature of the nation and its political structure are ongoing, but the data which follow in this chapter tend to support the former rather than the latter argument about incorporation.

NATURALIZATION AND ASSIMILATION

Citizenship is almost certainly the most significant dimension of assimilation outside of intermarriage. Political incorporation and involvement means creating new allegiances and becoming a part of a new society. Political incorporation is also at the heart of the way in which individuals and groups give voice to their concerns, and by extension it is the way in which they attempt to protect and enlarge their individual and group interests. Hence, political incorporation is a central part of the process of transforming immigrants to citizens, and in turn to participants. By becoming citizens, immigrants are following the path suggested by the initial attempts to create an American society with the motto *E Pluribus Unum*—"out of many one." Whatever the current debates about how new immigrants should be incorporated and the tensions and problems of blending groups from different backgrounds, it was clearly the intent of those who framed the U.S. Constitution to create a blended society.

Even though the motto remains a central part of American mythology and is imprinted on the coinage, the centrality of *E Pluribus Unum* is vigorously debated by those who see blending as insufficiently sensitive to the many cultures that are arriving during the current large-scale immigration. While questions of how the new immigrants were going to fit in were asked during the last great waves of immigration, the debate today is occurring in a different intellectual and cultural climate. However, even though it is true that the majority of the new citizens are often not English speaking and many have religions that are new to the United States, it is still true that the majority of the new immigrants are coming with the same motivations that brought the waves of immigrants at the onset of the 20th century. The desire for a better life, to make money, to give opportunities to their children, and to "move up" are still fundamental forces in the immigration process.

The current debates about immigration are not only about the size of the flows and about who should be admitted but also about the processes of incorporation, and more specifically about how citizenship should be extended and to whom—especially whether those who arrive illegally should be extended legal status. These and other questions have been the subject of both media commentary and academic debate, without any real consensus emerging. In general, most academic writing rejects the notion of assimilation, certainly of the "melting pot" invocation (Glazer, 1993; Gans, 1997), yet more popular writers argue that the melting pot can work again (Barone, 2001).[3]

Assimilation imagery is often translated, by the media and academics alike, into a smooth and linear process of succeeding generations gradually becoming American. But as the discussion in Chapter 1 showed, the assimilation process, even in the past and certainly today, is much more nuanced and complex than the mythic structures that have been erected around it. But at the same time it does not make it any the less relevant.

It is worth reiterating three points about the assimilation process—it is multidimensional, it is not irreversible, and it is conflictual; this is true in the current immigration context as it was in the past (Skerry, 1998). While we often speak about whether this group or that will assimilate, or whether one group or another is making more progress in assimilating, it cannot be reiterated too often that assimilation has economic, social, cultural, and political dimensions and that one does not neatly mesh with another. Progress on one dimension may not be followed by progress on another dimension. Skerry (1998) notes that Mexican Americans are intermarrying, moving to the suburbs, and electing Latino representatives, but at the same time many are having difficulty in making economic advances. Clearly, they have some of the dimensions of integration and assimilation but not all.

Assimilation, too, has its tensions. Skerry, using work by Olzak (1992) and Lipset and Raab (1970), argues that intergroup conflict is caused not by segregation per se but by occupational desegregation. It is when there is a weakening of the boundaries between groups that conflict emerges, not when one group is isolated from another. It is when individuals expand their opportunities and break beyond previously established group boundaries that conflict occurs. Greater interaction, rather than separation, leads to greater chances of conflict.[4] Tension and conflict arises especially when one group extends its power to create political advantage. Political progress for one group can have a divisive and nonassimilative impact for another group, which retreats into a fortress mentality. Thus, the extension of voting protections to Hispanics and Asians and the creation of special electoral districts to favor one or another ethnic group make sense in the short run, but to the extent that one group excludes another group the outcome can become divisive and create tension.

Moreover, tension can decline in one area and increase in another. Cain and Kiewiet (1986) point out that claims of economic discrimination decline from the first generation to the second and can decline further in succeeding generations, but the claims of social discrimination increase. This notion fits with the idea that Latino economic gains lead to an increase in social contacts and greater chances of friction and conflict. Certainly, there is evidence that economic advances translated into gains in the housing market have meant neighborhood transition in inner-city areas in Southern California, and consequent conflict between the black residents and Hispanic migrants who have moved into these neighborhoods. The influx of Latino households seeking rental and owner housing has sometimes brought conflict between previous African American residents who feel that their new neighbors are not sensitive to established patterns of behavior (Clark, 1998).

The differences between generations and their expectations are neatly summarized by the recent observations on language use by first, second, and later generations of immigrants. Many first- and second-generation immigrants have, as in the case of previous waves, adopted English and use it both in and outside the home. Now, many third-generation immigrants are concerned to "rediscover" the language of their grandparents and emphasize their cultural heritage. It seems that the challenge is not about assimilation and adaptation for the first and second generations, who are glad to blend in, but about the third and later generations who have higher expectations than their parents, and may not be so easily satisfied by material gains alone. Assimilation, once accepted, may become contested territory for the third generation. While some students entering the University of California, Berkeley felt that they were pretty much assimilated, others saw themselves as being "born again . . . at Berkeley," rediscovering their Mexican roots and heritage (The Diversity Project, 1991, quoted in Skerry, 1998).

It is not altogether clear how integration and assimilation will be played out in the coming decades. When new immigrants become a majority, as they are in California already, it is less meaningful to speak of assimilation, and "blending" may be a better term. In the end though, it is over time and through the children that the process will be defined, language changed, and citizens created.

THE CONTEXT OF ASSIMILATION AND CITIZENSHIP

As part of the discussion over assimilation and the creation of a forged or blended society, a vigorous debate has arisen over the nature of citizenship—and dual citizenship in particular. At least 92 countries now permit multiple citizenship for economic, political, or cultural reasons. Persons who enter from such countries are able to maintain and foster their ties with countries from which they have emigrated (Renshon, 2001). The advocates of dual citizenship see it as a step toward recognizing greater diversity and an increase in the idea of larger loyalties, rather than those narrowly focused on American interests. Many advocates of dual citizenship are also advocates of multiculturalism (Isbister, 1998). Indeed, the debates about dual citizenship are debates about loyalty and attachment. They are debates about whether new citizens will consider themselves part of their new society, will feel themselves American citizens. Will they be more loyal to the United States or Mexico, to the United States or Korea? An even more troubling issue to some is whether loyalty to religion will preclude loyalty to the state?

At the same time, as Renshon and others have pointed out, loyalty is a complex concept and an even more complex emotion. While it is often described as an attachment to a person, place, or thing, it can have a very wide range of intensities and meanings (Renshon, 2001). There is no question that immigrants are conflicted, whether or not they have dual citizenship. Immigrants sometimes hesitate to take out U.S. citizenship because they want to keep their ties to their homelands. At the same time, their children may be less conflicted, and that is the issue which is often lost in these debates. While first generation immigrants may look back fondly to their homelands, their children, growing up in a new society, are much less likely to have such feelings of nostalgia. In turn, some third-generation immigrants may be searching for lost ties and deeper cultural roots. The process is by no means straightforward or transparent.

There is particular concern about dual citizenship for Mexican nationals. In 1998 Mexico amended the constitution to allow Mexicans living abroad to have dual citizenship. This change in the way in which Mexico regarded longtime expatriates was designed to strengthen the links between its citizens living in the United States and their home country.[5] Does it matter? Renshon (2001) believes it does, and that such divided loyalty can—and may well—lead to conflicted affiliations. He contrasts the situation during previous waves of immigration, when immigrants were asked to learn the language, culture, and political practices of America, and to become a part of the fabric of American life, with the current situation when, if anything, cultural diversity and attachment to other cultures and societies are celebrated. Dual citizenship, with its associated bifurcation of attention and commitment, means that the situation may be ripe for local, if not national, conflict between citizens and immigrants. It seems clear already that some localities, and soon whole states, will have large numbers of citizens who retain strong connections to their societies of origin. If we accept the current numbers of some 8–10 million undocumented immigrants (Warren, 2000), then there are large numbers with diverse and perhaps conflicted attachments. The meaning of these connections has yet to be played out. While the connections may be potentially divisive, it is again the long-term evolution of the relationship which is important.

The issue of attachment and connection is embedded within the notion of trasnationalism. At one level, transnationalism has been used to describe all movements back and forth across borders, but as Glick-Schiller (1999) points out, this approach deprives transnationalism of its real meaning as encompassing the extent to which ties are maintained beyond the first generation and involve a notion of more than mere travel between nations. Transnational communities hint at links over generations and truly living in two worlds. Clearly, technological changes have increased linkages, and Dominicans in New York can easily board a plane to vote in Dominican elections and Dominican politicians can easily fly to New York for political activities (Foner, 2000). Even so, just how large and how strong the ties are is still a subject for debate. It is an open question too about whether the current ties of those who travel back and forth will continue for their children. Clearly, it is easier to travel now than it was a century ago, but there were transnational linkages then, too; it is surely not a new process (Foner, 2000). Incorporation occurred in the past and will likely continue to occur even as transnational migration and communities evolve and change.[6]

Such concerns as to a pluralistic society are certainly influencing our thinking about the evolution of a blended society, but it is worth reflecting that, as Barone (2001) and others observe, race is an arbitrary categorization. Where once native-born Americans referred to Irish, Jews, and Italians as "other" races, we now speak of Asians, Hispanics or Latinos, and Middle Easterners as "other" ethnic groups. But "Hispanic" is a Census term, while "Latino" is the increasingly popular term on the street and in the media; some Asians can speak Spanish as well as Chinese; and Middle Easterners span a wide range of ethnic and racial backgrounds, including white. The categories are more fluid than we might want to believe, and they will become even more so with increased intermarriage.[7] Thus, the several-decade view is likely to provide a more useful perspective than the short-term view, and with increasing multirace categories, racial categorization and population projections by race and ethnicity may no longer be central in the late 21st century (Ellis, 2000).

Moreover, the current ideas that assimilation is different today than in the past are also open to question. Although Portes and Zhou (1992, 1993) emphasize the current differences in the process of assimilation and even a return *from* assimilation, it may be the case that, despite the academic rhetoric, assimilation in the past was not fundamentally different than assimilation in the present. To reiterate my earlier arguments and those of Alba and Nee (1997), it is important to be wary about rejecting assimilation and to recognize that today as in the past the process is a fluid one, but with underlying continuities and eventual consequences for individuals, groups, and society. It was complex in the past, and it still is today.

One noteworthy obstacle to the continued incorporation of the new immigrants may be the changing attitudes toward earlier civil rights legislation (Barone, 2001). The civil rights movement and post–Vietnam War politics led to the adoption of legislation which established quotas and preferences to overcome the past effects of racism on African Americans and Native Americans. Whatever the logic of this approach, these preferences were extended to Asians and Hispanics as well, many of whom may not have experienced the same kind of discrimination as did African Americans and Native Americans. Now, we have the occasional (though usually unvoiced) concern that our elite universities may have "too many" Asians. Recent moves to confront the language of the Civil Rights Act of 1964 and to emphasize equality may lead to a more equitable society and to less worry about the numbers of one or another group in universities and the like. There is much to suggest that the new Americans of today, like the new Americans of the past, can be incorporated into American society, and that incorporation is essentially involved in becoming a part of the political process.

However, the central issues of this chapter concern what is happening with respect to citizenship. It is not the intent of this chapter to debate the nuances of citizenship or what constitutes American citizenship. The aim here is much simpler. True, the chapter is set within the context of assimilation, but the aims are at the same time more focused on understanding the rates of participation *as a measure* of adaptation to American society. The pivotal questions revolve around who are becoming citizens, how are they participating in their new society, and how are they influencing the new politics of America?

NATURALIZATION AND BECOMING AMERICAN

For the past two centuries or so naturalization has required a 5-year residency[8]; since 1952, with the McCarran–Walter Act, it has been open to all immigrants regardless of race, sex, or marital status.[9] Naturalization confers the right to vote, to work in some otherwise restricted government positions, to quality for a variety of federal benefits, and to be able to bring in some family members under current immigration law. These tangible benefits of naturalization are paralleled by less-tangible forces, including changing allegiances and being a part of a differently structured and democratic society.

Naturalization confers a status that is intended to be virtually indistinguishable from that of the native born. Naturalization is in some sense the end step in a process of developing loyalty to a new society and a new government. Survey data for Latino immigrants have documented that naturalized Latino immigrants are relatively quick to develop an attachment to the United States and to the political values that are at the core of the U.S. political process (de la Garza, 1992). An important finding of the surveys showed that naturalized citizens were much more likely to report allegiance to the United States than to their home countries (DeSipio, 1996a, 1996b).

The propensity to naturalize varies by country of origin, duration in the United States, and socioeconomic status. The relatively lower rates of naturalization for Mexican and Canadian immigrants is often explained by the proximity of these countries to the United States and the possibility of continued links across these adjacent borders. Formal education and income are also predictors of political participation (Wolfinger and Rosenstone, 1980). Manual laborers, often from backgrounds with modest education and often not intending to be permanent settlers, are less likely to become naturalized than are professionals who are likely to quickly change nationality (Portes and Rumbaut, 1996). At the same time, there is no question that professionals and entrepreneurs are likely to retain an interest in their former homelands, but they are also likely because of their educational levels to become involved in domestic affairs (Portes and Rumbaut, 1996). We might expect that the group that will become most politically integrated are the immigrant professionals who are likely to live away from the central-city immigrant concentrations and hence to be less identified with immigrants per se. The large numbers of Indian and other Asian scientists and professionals, are often relatively well integrated into their residential communities and, by extension, exhibit political adaptation as well. Many of these new residents of communities in the San Francisco metropolitan region are active in local school districts and members of a wide range of community organizations. They are willing to step forward to participate at the grassroots levels, and some have the notion that eventually they may become more active in city and state politics.

Politics, Immigrants, and Citizenship

Participation in the political life of the United States occurs through individual voting by naturalized citizens, but also through the voting structures created to elect representatives to the state legislatures and the U.S. congress. On the one hand, individuals express their preferences by voting for candidates of one of the political parties, and on the other they can, as members of particular ethnic groups, participate in actions to increase the power base of the ethnic group. Participation in "group politics" was the way in which early immigrant groups, notably the Irish and the Italians, expanded their political base, and created paths for their members within the political power structures. Now we see this same process with a diverse range of new immigrants. The earlier discussion of the election to the Minnesota Senate of Ms. Mee Moua, a Hmong immigrant from the Laotian highlands (*New York Times*, February 2, 2002, p. A13), parallels the way in which Irish, Italian, and German immigrants made their way in U.S. political society some 80 years ago: "After 25 years, the community is finally engaged in the mainstream . . . we are no longer visitors to the county" (*New York Times*, February 2, 2002, p. A13). A similar—though perhaps more nuanced—story can be told for Caribbean immigrants in New York (Kasinitz, 1992). The outcomes may be contextually different, but they show the same pattern of activism, community participation, and engagement.

While routine electoral politics is about electing individual candidates, the Voting Rights Act created new ways in which ethnic groups, including new immigrants, can increase their voice in the political process. Because immigrants, documented or undocumented, are counted in the U.S. Census they do impact the way in which voting districts are drawn. Congressional districts and state legislative districts are apportioned on the basis of all persons, not just the adult citizens who constitute potential voters. Because of this requirement, and the U.S. Supreme Court and other federal court rulings based on the Voting Rights Act, some districts have been drawn to enhance the election of ethnic minorities to state legislatures or to the U.S. Congress. The outcome is sometimes districts with many fewer voters than might be expected from the usual relationship between the size of the district population and the number of voters.[10]

Strengthening minority representation has created a sea change in political participation and outcomes. As of this writing, there are 27 members of Congress who are Latino or Asian (Member Profile Report, Lexis Nexis Congressional Online Service. Bethesda, Maryland: Congressional

Information Service.). There were approximately 199 elected or appointed state officers who were Hispanic in the early 1990s (Reddy, 1993), and another 386 Asian elected or appointed representatives of Asian backgrounds (Gall and Gall, 1993). Almost certainly these numbers have increased in the past decade, and there are large numbers of foreign-born representatives on city councils, school boards and other governing bodies of U.S. municipal areas. The increase in representation will continue as the foreign born naturalize and enter local and national politics. In general more political participation is seen as a positive outcome, and one which serves to bind the new citizens into the civic life of the new society. Greater citizen participation in a liberal democracy contributes to the free expression of competing voices through democratic deliberation (Phillips, 1994; Gutmann and Thompson, 1996). The basic arguments in favor of more participation, as outlined by political scholars, is that greater participation of citizens means that more individual opinions and suggestions are being considered, which in turn can influence the outcomes. Greater individual participation translates into more effective representation, thereby producing a collective outcome that is closer to the common will.

Such participation is valued, too, for the growth it brings to individuals. As citizens become part of the political process, they increase their connections to the community and develop into members of the democratic body politic. Greater participation also brings legitimacy to leaders and provides popular support for particular political outcomes. Greater citizen participation in elections signifies the concern of the population and expresses their wish to be heard about who will govern and with what policies. If more political participation is desirable, then traditionally disadvantaged and underrepresented groups who participate can create public policies which take their concerns into account.[11]

In general, the notion of more participation has been invoked positively. Citizenship drives and calls for greater participation in the election process are seen as ways of increasing political activity on the part of new immigrants. At the same time there are those who raise questions about "winner-take-all" elections and suggest the need for alternate forms of the electoral process, including proportional representation (Guinier, 1993). Changing the electoral process to allow proportional representation would further enhance the voices of underrepresented minorities. Whether or not the form of participatory democracy will change, more participation will lead to greater involvement of minorities and more voices, and it is the professional and educational immigrant elites who will first transform the political process.

NATURALIZATION RATES AND EXPLANATIONS

There has been a sea change in the size, nature, and timing of naturalization in the past decade. The numbers have increased, the process involves immigrants from many more ethnic backgrounds, and there is evidence that immigrants begin the naturalization process as soon as they are eligible. During the 1970s and 1980s, about 200,000 persons a year became naturalized citizens. That number is now more than three times as large and is moving closer to a million naturalizations a year.[12] The composition of the new citizens is no longer European but a wide mix of ethnic origins and races. Recent data show that less than 10% of recent naturalizations are from Western Europe. Now Asia and Central and South America are the regions with the greatest number of immigrants becoming U.S. citizens. Finally, the limited evidence suggests that a significant number of immigrants are applying for citizenship as soon as they have fulfilled the 5-year residency requirement (or 3 years in the case of foreign-born spouses married to an American citizen). A comparison of naturalization rates for cohorts admitted in 1977 and 1982 shows that their naturalization rates are 5.3 and 7.5% (U.S. Immigration and Naturalization Service, 1996).[13] Although the rates for those who file for and accomplish naturalization as soon as possible is relatively low, the increase in naturalization rates is consistent with the recent media-reported jump in applications for citizenship. Some of this is a response to the fear related to the negative campaigning on Proposition 187 in California, and other responses may include the recent changes in Mexican law which allows dual citizenship.

As of the late 1990s, about 37% of all foreign-born persons were naturalized.[14] This amounted to some 10.8 million of the approximately 30 million foreign-born persons then residing in the United States. Because of the current debate over who and how many immigrants should be admitted and the discussions of who becomes citizens, it is useful to have some broad outlines of the naturalization process and its outcomes.

Overall, the findings are consistent with other studies of naturalization. Rates are higher for those who have been here longer, and they increase with higher levels of education, professional occupations, and income. Because speaking English is a part of the test for citizenship, it is not surprising to find that those with higher English-language skills are more likely to be citizens (Bouvier, 1996). That earlier arrivals are more likely to be citizens is a finding consistent with the argument that incorporation takes time. For all of the foreign born who arrived before 1970, before the recent large-scale immigrant flows, the rates are at or above

80% (Figure 6.1). This, certainly reflects a level of incorporation which suggests that immigrants in the latter part of the 20th century are replicating the assimilation and political incorporation of earlier immigrant generations. The rates of naturalization decline across all ethnic groups, with the most recent arrivals at very low levels of naturalization—again an expected outcome, as the minimum time for naturalization is 5 years. The discussion of incorporation and naturalization often focuses on the 30–40% of immigrants who have been resident in the United States for 20 years and are still not citizens. Two corollaries are important. First, we know that a large number of the foreign born are undocumented, and the naturalization rate is computed for all foreign born measured in the U.S. Census, not just legal entrants. Second, for many immigrants it may not be that they are unwilling, but rather that they are uncomfortable in English-language skills, or lack even the basic skills or the limited funds to complete the citizenship process. It is true that Hispanics in general have lower rates of naturalization, and it is this group which is often the focus of questions about assimilation and incorporation. I further explore these issues later in this chapter.

The likelihood of naturalization increases with age, and older immi-

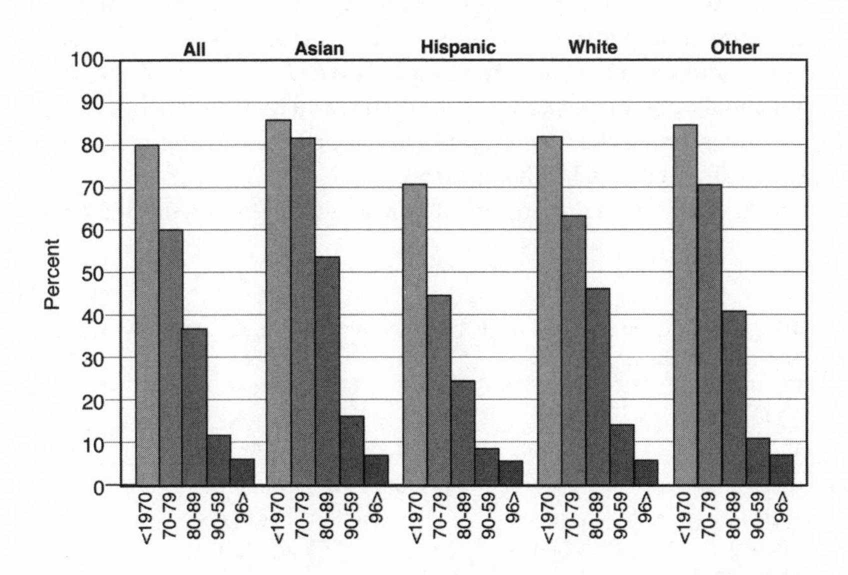

FIGURE 6.1. Naturalization levels in the United States by duration and ethnic background. *Source:* U.S. Bureau of the Census, Current Population Survey, Combined File, 1998–2000.

grants who arrive in later periods are more likely to become naturalized than younger immigrants of the same period. Those older than 45 years are twice as likely as younger immigrants to be naturalized (Table 6.1). Again, this is not an unexpected finding. Older immigrants are likely to have families, and perhaps to have a greater incentive to complete the citizenship process.

Education and occupation have distinct correlations with the likelihood of naturalization. Those with a college education are twice as likely as the "less than high school" group to be naturalized (Table 6.2). The critical dividing line is "more than a high school education," as is shown clearly in Table 6.2. This finding in part helps us understand the low rates of naturalization for Mexican immigrants, who almost universally have relatively low levels of education if they came to the United States as adults. Education and occupation are, of course, intertwined. Thus, we find that professional and white-collar workers are more likely to be naturalized than are farmworkers, but even here there are important ethnic differences (Table 6.3). While Hispanic farmworkers, largely made up of Mexican and Central American immigrants, are very unlikely to naturalize, professionals who are Hispanics, including Mexican and Central American immigrants, are nearly five times as likely to be citizens. There is a major break in the levels of naturalization between professional and other white-collar workers and less-skilled immigrants. The implications of these findings are relatively straightforward. Older immigrants and professionals are very likely to want to participate in their new society, whereas less-skilled immigrants are simply struggling to get ahead economically—citizenship has to wait.

As is consistent with arguments presented elsewhere in this book, it

TABLE 6.1. Naturalization Levels by Age and Year of Arrival of Immigrants in the United States

	Age			
Year of arrival	14 years	15–24 years	25–44 years	45+ years
1996	8.7	3.4	4.0	8.2
1990–1996	10.4	7.6	10.8	17.3
1980–1989	30.1	31.0	36.3	41.2
1970–1979	—	52.5	57.8	61.6
< 1970	—	—	77.2	79.5

Source: U.S. Bureau of the Census, Current Population Survey, Combined File, 1998–2000.

TABLE 6.2. Naturalization Rates by Educational Attainment of Immigrants

Education	Naturalization rate
< High school	25.2
High School	40.8
Some college	46.9
College	50.3

Source: U.S. Bureau of the Census, Current Population Survey, Combined File, 1998–2000.

is middle-income (middle-class) immigrants who are most likely to assimilate and participate. An examination of the combination of occupational levels and middle-income levels, defined as those with incomes in the range of approximately $33,000 to $84,000,[15] provides an important window on the rates of naturalization (Table 6.4). Naturalization rates are significantly higher for upper-middle-income immigrants, particularly for Hispanic immigrants, but for all groups there is a steady increase in naturalization rates with increasing incomes. When we examine the rates by income, for professional and white-collar occupations, the rates increase again. The increases are most notable for the Hispanic foreign-born population, where the naturalization rates increase by nearly 20% over those for income alone (Table 6.4, panels b and c). There are some interesting subpatterns in the naturalization rates when we compare the rates for those in professional occupations and those in white-collar occupations. Asian foreign-born persons in white-collar occupations are much more likely to be naturalized citizens than are those in professional occupa-

TABLE 6.3. Naturalization Rates by Race, Ethnicity, and Occupational Status of Immigrants

	All	White	Hispanic	Asian	Other
Professional	51.7	50.8	52.8	51.0	56.6
White-collar worker	47.7	50.0	61.3	55.6	36.2
Service	30.2	38.9	21.9	40.1	41.0
Blue-collar worker	30.1	42.9	21.9	45.9	41.7
Farmworker	14.3	23.7	12.2	45.4	36.6

Source: U.S. Bureau of the Census, Current Population Survey, Combined File, 1998–2000.

TABLE 6.4. Naturalization Rates by Economic Status, Ethnic Origin, and Occupational Status of Immigrants

	Ethnic Origin			
Economic status	White	Hispanic	Asian	Other
a. All occupations				
Lower	48.6	19.7	33.4	33.5
Lower-middle	48.3	24.6	42.0	39.7
Upper-middle	50.8	36.6	48.7	44.3
Upper	49.2	42.5	54.1	53.7
b. Professional occupations				
Lower	39.0	33.1	32.0	43.1
Lower-middle	44.7	48.0	36.3	54.9
Upper-middle	53.3	55.8	49.8	52.5
Upper	55.4	69.4	60.8	52.5
c. White collar occupations				
Lower	43.0	28.7	42.3	37.5
Lower-middle	52.6	37.4	52.0	30.7
Upper-middle	57.1	54.6	59.3	34.3
Upper	47.3	59.7	64.7	34.3

Source: U.S. Bureau of the Census, Current Population Survey, Combined File, 1998–2000.

tions. Basically, the reverse is true for Hispanic immigrants. The explanation almost certainly resides in the nature of the professional and white-collar occupations of each group. Asian immigrants, as was clear in the discussion of occupations in Chapter 4, are quite likely to be in classic white-collar occupations such as laboratory-technology occupations. But still these are occupations that do not require the same educational levels as the professional occupations. This leaves unanswered the question of why Asian professionals have lower rates than their white-collar compatriots. A twofold explanation rests on the following arguments: (1) that at least some Asian professionals may still consider returning to their home countries, and (2) that at least some of these professionals may have working relations with their home nations. Hence there is a lower probability of naturalization.

For some specific groups, the combinations of income and occupation produce very high levels of naturalization. Three examples illustrate the outcomes for particular ethnic origins. Cubans who are foreign born have the very highest naturalization rates despite their oft-cited rhetoric about someday returning to Cuba. More than 70% of upper-middle-income foreign-born Cuban immigrants are naturalized, as are more than 90% of upper-middle-income Cuban professionals. No other group has such a high level of naturalization, but Korean and Middle Eastern foreign-born upper-middle-income professionals have naturalization rates of 83.8 and 72.8%, respectively. The rates for Mexican upper-middle-income individuals are lower: only about 28% are naturalized, but nearly half (48.4%) of white-collar Mexican foreign-born individuals are naturalized. These specific group rates illustrate the divergence among groups, but they also reinforce the sensitivity of the process of naturalization to socioeconomic status.

In an earlier chapter I made the argument that many of the social integration processes were overlapping and that when naturalization and homeownership are interrelated the nature of the reinforcing processes are extremely clear. Homeownership and naturalization go hand in hand. Ownership rates are much higher for naturalized citizens and increase dramatically with age (Figure 6.2). Even Hispanic households, who have the lowest homeownership rates overall, are very likely to be citizens. The homeownership rates for other naturalized groups are higher.

Overall, some groups are slow to naturalize, but middle-income professional and white-collar immigrants are naturalizing at relatively high rates. And there is a close relationship with homeownership, too. It is not unreasonable to argue that the interaction of income, education, and naturalization serve as markers of political assimilation. One must assume that even though the same process may not be working for groups with lower levels of education and income, their children, born as citizens, will be much more likely to participate in the electoral process. Some data on second- and third-generation participation will address this question later in the chapter.

Becoming a Voter: Political Participation

Naturalization is the initial step to political participation, and political participation, like naturalization, is linked to duration in the country, education, occupation, and income.

The national voter turnout (the percentage of voters who actually cast a ballot) in the 2000 presidential election was 51% nationally and approximately 60% of the native-born population. Naturalized citizens who have been in the United States since before 1980 participated at or above the national rate; the more recent arrivals participated at lower rates, pos-

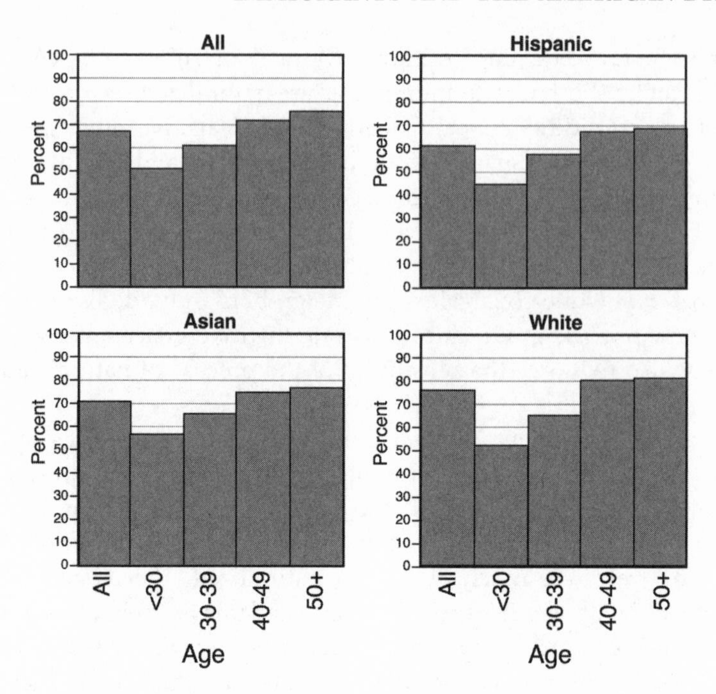

FIGURE 6.2. Homeownership rates for naturalized U.S. citizens. *Source:* U.S. Bureau of the Census, Current Population Survey, Combined File, 1998–2000.

sibly reflecting the typically lower rates among younger adults generally. Indeed, we might expect the most recent arrivals who have become citizens to participate at higher rates, or at least at the same rates, as earlier arrivals (Figure 6.3). The decline does reflect an age effect—people under 35 versus older adults. However, we can also interpret the decline in participation in the same manner as our interpretation of the falloff in naturalization. Longer-term citizens show a greater involvement in their communities and hence a greater likelihood of voting.

Older naturalized citizens vote at or above the national rate, and those with some college education or with college degrees also vote at or above the national rate. Those in professional occupations and, by extension, those with high incomes also vote at rates which are above the national levels (Figure 6.4). Again, as in the naturalization rates, there are marked differences across ethnic groups. Canadians, Cubans, and Central Americans have the highest levels of voter participation. While citizens of Middle Eastern origin are highly likely to naturalize, they vote at rates

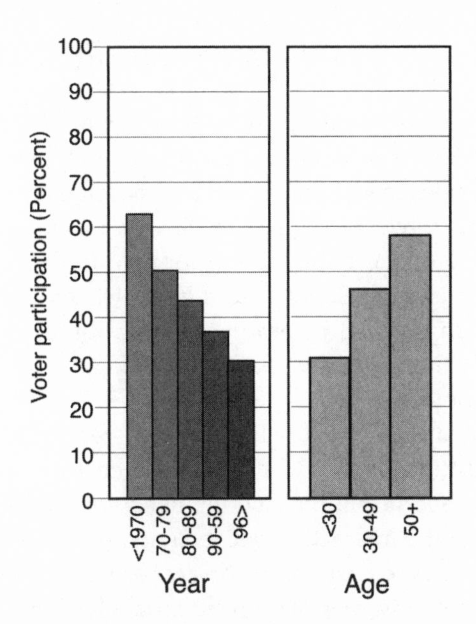

FIGURE 6.3. Voting participation by period of entry to the United Sates and age (percentage of foreign born who voted in the November 2000 election). *Source:* U.S. Bureau of the Census, Current Population Survey, Voter Supplement File, November 2000.

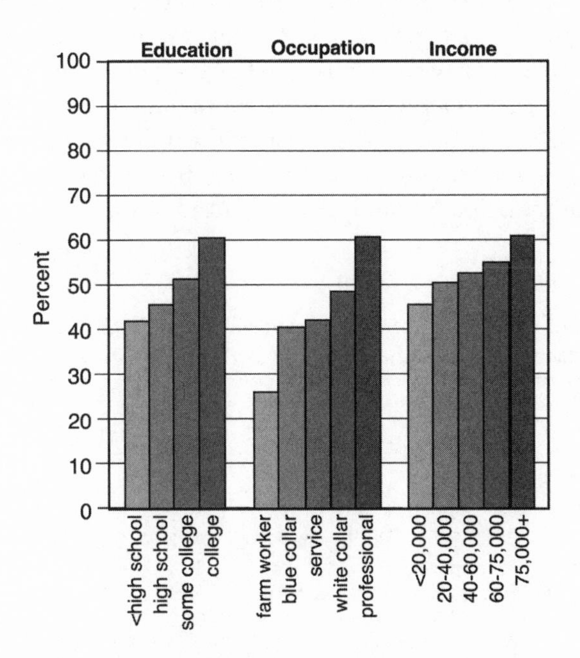

FIGURE 6.4. Voting participation of naturalized U.S. citizens by education, occupation, and income. *Source:* U.S. Bureau of the Census, Current Population Survey, Voter Supplement File, November 2000.

that are similar to those of naturalized citizens from Mexico: about 42% in each case (Table 6.5). The data suggest that once Mexican-origin immigrants make the decision to naturalize and become citizens, they are as active participants in elections as other foreign-origin groups. Yet, more than half of many of the naturalized groups did not participate in the 2000 presidential election. Getting out the immigrant vote has become an issue within the immigrant communities, as political organizers recognize that it is only through participation that the new citizens will be able to influence the political process.

Homeowners are much more likely to register and also to vote. They are more than 10% more likely to be registered than renters and have about the same difference in their voting behavior (Table 6.6). One of the notable findings of the interaction between homeownership and voting is that nearly two-thirds of Hispanics who are owners are registered to vote, and nearly 56% voted in the 2000 presidential election.

The report on voting in this chapter, drawn from the 2000 presidential election, is representative of the involvement of the immigrant community translated into citizen participation. The fit is not perfect, but the evidence suggests that naturalized citizens are not too different from the native born in their participation rates. Moreover, given their first-generation unfamiliarity with a new political system, the rates of participation in the electoral process are encouraging data on the levels of involvement of new citizens. The question that follows is how this participation changes in later generations (see also Bass and Casper, 2002, for a discussion of participation and length of residence).

TABLE 6.5. Political Participation (Percent Voted) in the 2000 Presidential Election by Place of Birth of the Naturalized Foreign Born

Hispanic	Percent voted	Asian	Percent voted	White	Percent voted
Mexico	40.5	China	38.8	Europe/Russia	57.5
Central America	59.5	Korea	53.1	Middle East	41.9
Cuba	65.7	Philippines	46.7	Canada	66.8
South America	56.2	Asia/India	46.4	Other	52.4
Dominican Republic	43.2	Southeast Asia	50.1		
Other	54.3	Japan	55.0		
		Other	39.9		

Source: U.S. Bureau of the Census, Current Population Survey, Voter Supplement File, November 2000.

TABLE 6.6. Registration and Voting by Homeownership

	Voted		Registered to vote	
	Owner	Rental	Owner	Rental
All foreign born	62.1	48.3	54.4	41.1
Asian	53.1	47.4	44.3	41.1
Hispanic	64.2	45.0	55.9	37.3
White	66.5	52.5	59.3	43.5

Source: U.S. Bureau of the Census, Current Population Survey, Voter Supplement File, November 2000.

The Second Generation and Later

The proportion of immigrants eligible to participate in the electoral process will increase in the coming decades, and the children of immigrants who are citizens by birth will also become a major factor in the electoral process. These second-generation voters will play an important role in shaping the electoral landscape. Already in 2000, second-generation adults accounted for more than 8% of the electorate, and this group will grow and gradually replace the second generation from Europe. The electorate will have a different look in the coming decades, as the children of late-20th-century immigrants become adults. The significance of the second generation has not been lost on local or national politicians. During the recent elections in 2000 there were repeated calls by ethnic organizations urging their members to participate in the electoral process.

There is, in fact, only limited research on immigrant political participation and even less on immigrant political incorporation. Beyond the general statements that voting by immigrants is influenced by the length of stay (Uhlaner, Cain, and Kiewiet, 1989; Cho, 1999), by their ability to speak English, and that overall they are less likely to participate than the native born (Junn, 1999), there are only limited national-level studies of voting participation. There are even fewer studies of the role of generation status. Yet, it is this latter which will play an important role in changing the face of the American electorate. From my perspective here, I am more concerned with participation as an indicator of immigrant incorporation and assimilation than of the political outcomes of the changing electorate.

At least one study has suggested that the process of political incorporation does not square well with our ideas of assimilation, at least the linear conceptualization of assimilation. Ramakrishnan and Espenshade (2000) argue that the differences in participation across immigrant generations casts doubt on the standard assimilationist accounts of political incorporation. They show that the second generation is the generation more likely to vote at a higher rate than later and succeeding generations. But it may be that this is less of an argument against political incorporation than it seems. Just because participation is lower, it does not mean lower levels of assimilation. In some sense the later generations are perhaps behaving more like the native born, who are less likely to participate in voting than are new immigrants. True, the issues are complex, as they are involved with the secular decline in voting participation in general, but overall the evidence tends to support rather than undermine a view of incorporation.

Other research shows that there is a progressive increase in voting participation as immigrants spend more time in the United States (Ramakrishnan, 2001). The pattern holds across different ethnic groups, and even though there are generational differences and differences across ethnic groups there is still a real increase in participation for two of the three groups analyzed (Figure 6.5). For Asian citizens, the turnout rates climb steadily, and are highest for the third generation. For foreign-born whites voting participation peaks in the second generation, and there is a slight decline in later generation respondents. In this classic interpretation, voting increases over time and generations. Hispanics do not show generational gains. Participation increases with age for the first generation, but later generations do not participate at increased rates. However, it is important to put this decrease or (as I prefer) stabilization, of the participation rates into perspective. Despite the falloff of the second and third generations, participation rates are substantially higher than younger first-generation voting rates. Ramakrishnan (2001) interprets these results as a lack of support for assimilationist perspectives. Certainly, for Asians, they can be viewed as supporting political incorporation, and perhaps for whites, too. Hispanic groups are likely to continue to change now that Mexicans can become U.S. citizens without giving up their Mexican nationality.

An alternative and reasonable interpretation of the graphs is that they support behavior more like the native born. It is true that Latinos and Asians have faced institutional barriers to participation, most of which are now removed. It is also true that language plays a role in levels of participation, as does the home-country political context of the immi-

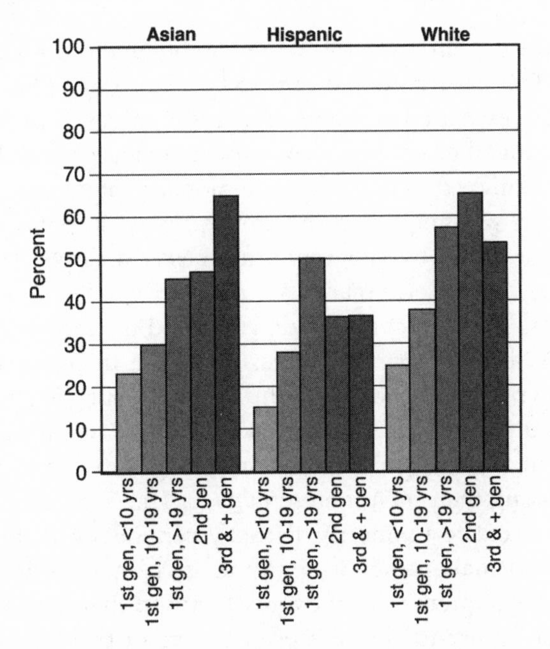

FIGURE 6.5. Generational differences in voting participation, 1994–1998. Redrawn from Ramakrishnan, 2001. *Date source:* U.S. Bureau of the Census, Current Population Survey, November 1994 and November 1998.

grant populations. Although it is hard to assess such factors, a political context in which democratic elections were not possible could have an impact on political participation in the United States.[16] Certainly the participation levels of the naturalized foreign born are lower than those of the native born, but the evidence presented later suggests that this may be a temporal rather than a generational outcome.

POLITICAL REPRESENTATION, IMMIGRANT ENCLAVES, AND THE POLITICS OF SPACE

Up to this point, the discussion has been about individual incorporation, evidenced by naturalization and its translation into voting participation. It is also important to see the results of voting in their aggregate outcomes, and in a spatial context as well. It is almost a truism to say that the opportunities for political activity and involvement are shaped locally (DeSipio, 1996a). Participation occurs in specific geographic contexts and is played out in increased numbers of representatives from ethnic groups. People living together tend to be of similar socioeconomic backgrounds and to share similar concerns. Their commonality of interest, which arises from residential propinquity, is in turn expressed in political choices by voters who are linked by geographically based districts. Thus, interests and voices are tied to local places.

The Voting Rights Act has made place and the incorporation of ethnic minorities, a central part of increasing the participation of minorities in the political process. Federal law, through the Voting Rights Act of 1965 and its amendments of 1970, 1975, and 1982, has made it possible to create districts in which "protected" minority groups have an increased chance of electing their own representatives. The *Gingles* criterion makes it mandatory to create a minority district when the resident population is large enough to make up 50% of any district. In 1968, in *Thornburg v. Gingles,* the U.S. Supreme Court expressly found that African Americans must have the ability to elect a representative of their choice once they were sufficiently numerous among the population (more than 50%) to do so. In urban contexts, concentrations of African Americans had been disenfranchised by splitting their numbers among several districts so that their political power was reduced. Creating districts in which African Americans could be elected was thus a political response to past discriminatory practices. Now the process has been extended to creating "ethnic districts" for other protected groups, often made up of recent migrants. There are contrasting scenarios for empowering or creating *influence* districts. In the past a resident concentration of Latino voters was sometimes split among several districts to dilute their power. In empowering or influencing scenarios Latinos, say, can be concentrated in one district, which would maximize their chance to elect a Latino representative, or the Latino population could be divided into two districts where their "influence" might make it possible to elect two Latino representatives. A more extended discussion of the issue of dominance and influence districts can be found in Clark and Morrison (1995). It is this spatially based process that marks the extension of individual incorporation and which leads to the increase in electoral representation.

A study of parity scores, the percentage of a group's elected officials as a percentage of that group's percentage of the total population, suggests that these changes have impacted electoral outcomes. A specific study of Hispanic participation shows a steady trend of increased electoral presence beginning in the early 1970s (Rosenfeld, 1998). Hispanic electoralism, and more specifically Mexican American electoralism in the Southwestern states, especially Texas, Arizona, New Mexico, and California, increased from parity scores of about .2 to consistently above .5 and at or near 1.0 in some states (Figure 6.6). The parity scores[17] measure the extent to which an ethnic group, in the present case, Hispanics, is increasing its participation by electing more of its own representatives. Another interpretation is to suggest that the parity scores measure the way in which the Mexican American community was able to translate its new voting power into elected representatives. It is reasonable to argue that

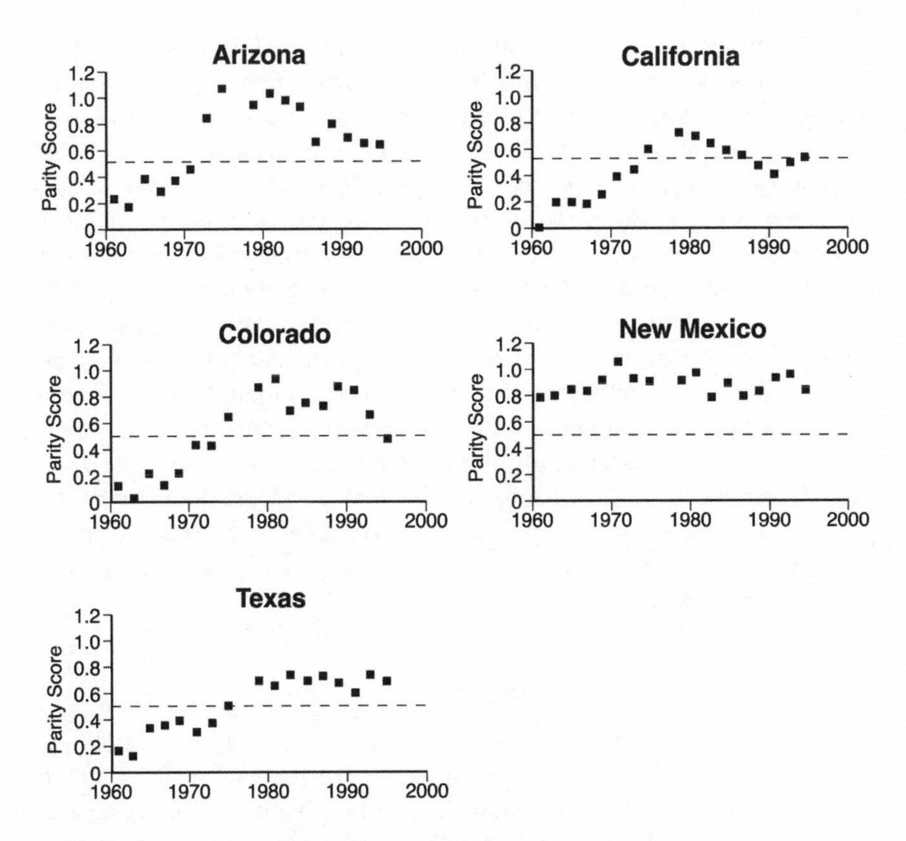

FIGURE 6.6. Growing Hispanic parity index for Spanish-surnamed elected officials in the Southwestern United States, 1960–1995. *Source:* Redrawn with permission from Rosenfield, 1998.

the process was aided by the special protection of the Voting Rights Act, which extended the idea of ethnically cohesive districts from African Americans to Hispanics.[18] Even so, while creating special districts in which Hispanics could be elected was important, ethnic politics also played a part.

The achievement of political power over time socializes immigrants into the functioning of mainstream institutions and gives them the voice to feel that they are part of those institutions (Portes and Stepick, 1993). NALEO (the National Association of Latino Elected and Appointed Officials) and SWVRP (the Southwest Voter Registration Project) were founded in the late 1970s and were directly aimed at promoting civic and electoral participation of Mexican Americans (see NALEO Educational Fund, 1989, 1992).

The implications for the political process and political incorporation are more complicated for new immigrant citizens than for the previous redress of past discriminatory practices against African Americans. Two issues to be resolved are whether immigrant citizens need to be specially empowered and what are the effects in pluralistic locales. In such locales, the procedure may not be so simple and may lead to outcomes that are problematic, or even counter-productive to the long-run incorporation of immigrants. One group's degree of spatial concentration, relative to others in the population, will determine how easily a single-member electoral district can be created to recognize a particular minority. For the often highly concentrated African American population it has been relatively straightforward to empower such groups. For Asians, who are often quite scattered residentially, it has been much less feasible. For Latinos, the process is easier in some locales than in others. Sometimes the process requires a certain amount of cartographic contortion to bring about a district, and empowering a group may do violence to other communities of interest.[19]

Space, Place, and Assimilation

Geography matters, and nowhere more so than in redrawing the political map to reflect the changing demography of cities and states. To make the issue concrete, it is useful to take up a recent exchange in Southern California where pluralistic locales are more common than otherwise nationally. However, it is clear that the context in California is a harbinger of the process that is likely to occur across the nation, as the districting process continues in the 21st century. It is useful to recall the previous discussions of the increasing dispersion of immigrants throughout the towns and cities of the United States as we examine what is happening in particular places in response to the changing demography of immigration.

In general, redistricting tends to protect incumbents, Democratic and Republican alike. In California, the Democratically controlled legislature protected Democratic incumbents and increased by one the number of Hispanic seats. However, the redistricting, illustrated in Figure 6.7, moved some Latino voters from a congressional district in which they were relatively concentrated to an adjacent district where their numbers were not as large. The current congressional district in the San Fernando Valley, District 26, encompasses a large area in which Hispanics are a near majority. The new District 28 substantially decreases the Hispanic presence in the new district. Two different perspectives on the change in districts are contained in a recent debate about their demographic composition and the implications for representing Latinos.

The redistricting generated a lawsuit by the Mexican American Legal Defense and Educational Fund (MALDEF) alleging that the redistricting discriminates against Latino voters. In response to the suit, two Latino legislators commented publicly on the issue of discrimination, and in so doing raised just the issue of influence or dominance in voting districts and the way place is translated into political influence. The suit claims that the new district, Congressional District 28, does not have a sufficient number of Latino voters; in other words, it is not a "safe" seat. The Latino legislators squarely face the issue that will become increasingly salient in issues of political participation and political influence. The legislators argued that the MALDEF lawsuit is racially divisive, suggesting that "more and more, California is reaping the benefits of multiracial coalitions. The voice of Latinos in California is stronger because electoral politics and issues are no longer just about race" (*Los Angeles Times*, November 1, 2001, p. B13). These legislators made two points which are relevant to the discussions of political influence and power: first, that Latino candidates do not require a district packed with Latino voters; second, that Latino success was based on proclaiming that the Latino agenda was the American agenda. Thus, place is more important than race. The legislators argued that most citizens vote for the most qualified candidate regardless of race or gender. While the latter may not be a total truth, it strikes at the issue which must be central in a reconstituted electorate.

The growing pressure to recognize the changing composition of the nation's population has created a concern with what are called "communities of interest," including how to create the greater empowerment of

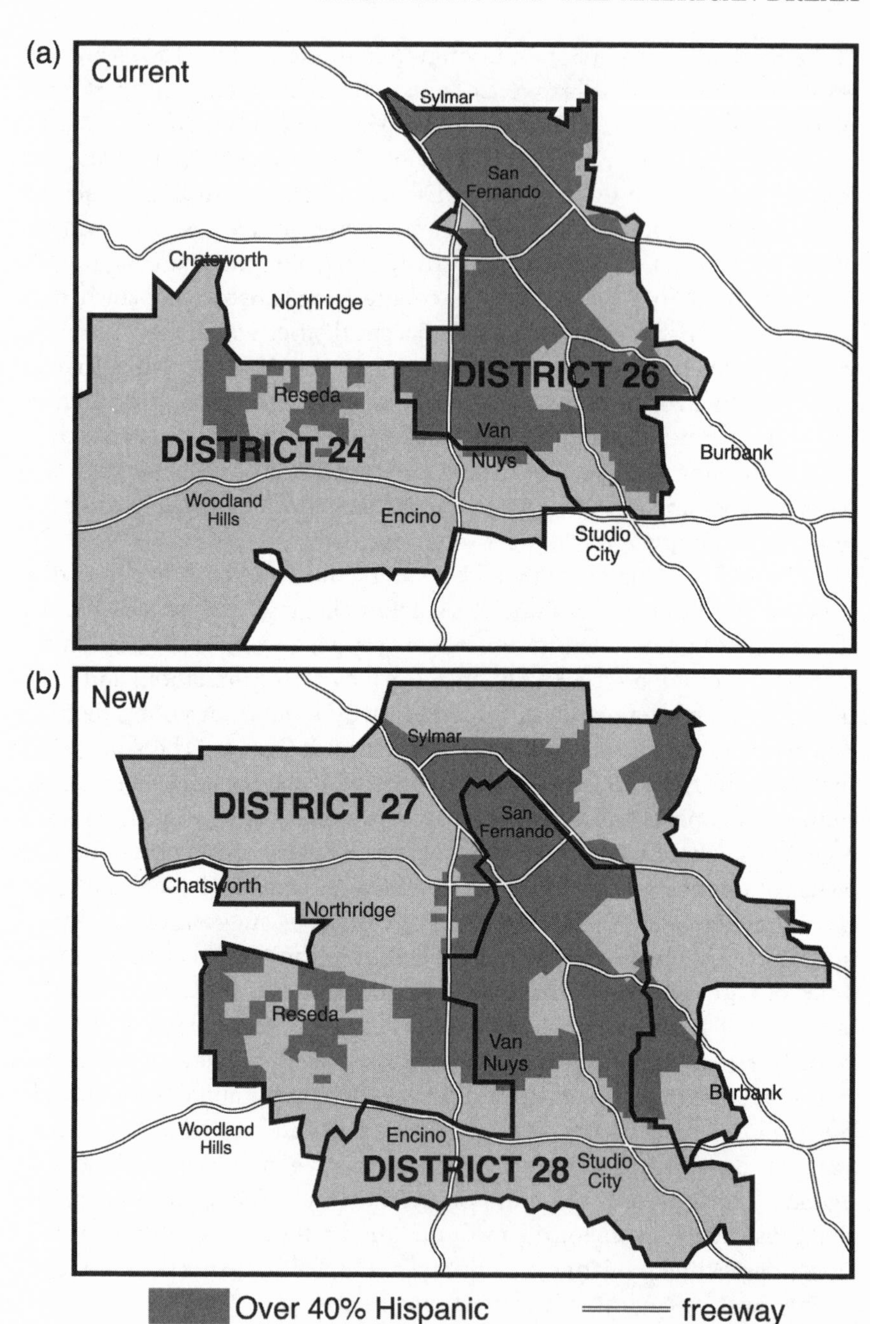

FIGURE 6.7. Creating congressional districts and protecting minority (ethnic) voters. *Source:* Redrawn from California Legislature, California Supreme Court Special Masters, *Los Angeles Times,* November 5, 2001, p. B2.

groups (Clark and Morrison, 1995); such communities of interest are, in essence, concentrations of people living in a specific geographic location and with common economic, social, and political interests. The problem in creating special voting districts comes about because communities of interest often overlap with ethnic or racial identity, and the question is how to enhance one without simultaneously weakening another. In Southern California, an attempt to reconfigure the voting districts in a school district may not enhance the voting strength of both African Americans and Hispanics at the same time (Clark and Morrison, 1995). As the mix of ethnicities in urban areas increases, the problem of districting will also increase.

The question remains whether districts should be configured to provide empowerment through "safe seats" or through influence districts. As the nation's cities experience additional demographic changes, and as the numbers of people of different ethnic origins enter the large cities in particular, the path of empowerment will increasingly bring into stark contrast the two possibilities of melting pot assimilation, on the one hand, or ethnic assertiveness, on the other. This is an important issue that goes to the heart of how the structure of political influence will evolve.

OBSERVATIONS AND CONCLUSIONS

Naturalization and political participation increase with time. They are markers of the way in which immigrants become part of their new society. This is not to argue that the process is simply linear or its pace is uniform for each group, or that it is without hitches and problems, only that the evidence seems to favor an ongoing process of change in which previous noncitizens become citizens and then political participants. In general, immigrants do not participate at any greater rate than the population as a whole; in fact, they vote at a slightly lower rate. However, just as for the native-born population, the rates of participation increase with age, education, and more professional occupations. The evidence from this chapter, like the evidence in the earlier chapters, emphasizes the role that the immigrant middle class plays in reforming the structure of a blended society. The idea, even if still in its infancy, that race-based politics is not the best process for producing such a blended society lends further support in favor of blending and assimilation. The question, which must now be reevaluated, is the extent and growth of the new immigrant middle class, as it is this group which will be at the forefront of the changes that have been considered in this chapter.

NOTES

1. The term "alien" is used by the U.S. Immigration and Naturalization Service for those who hold "registration cards," the so-called green cards which signify legal admission to the United States.
2. Although use of the term "Americanization" is often viewed as problematic because of nativist associations during the last great wave of immigration in the early 20th century, it is a term used most recently by the U.S. Commission on Immigration.
3. This is not to argue that the "melting pot" process worked in a systematic fashion at any time in the past. To the extent that we can invoke the melting pot, we are invoking an imperfect process that managed to include a wide variety of immigrants over a relatively long period of time and create an American society. Clearly, arguments about segmented or partial assimilation are attempts to provide this more nuanced understanding of assimilation.
4. Others have made similar points for specific groups like the Jews (Higham, 1975).
5. Renshon (2001) makes a strong case for a Mexican economic motivation in creating dual citizenship. He argues that the immigration northward is not only an economic safety valve, it is also a direct benefit to the Mexican economy as the immigrant population transfers somewhat on the order of U.S. $8 billion in remittances annually, far more than the total amount of official U.S. foreign aid (*Los Angeles Times,* September 16, 2002, p. C3).
6. Both Foner (2000) and Glick-Schiller (1999) examine the complexities of transnationalism—its implications for political incorporation, the evolution of the nation-state, and the personal narratives of migrants.
7. For a discussion of intermarriage and its implications see Farley (1999) and Perlman (1997). Both authors suggest that intermarriage will increase and will fundamentally change the way in which we researchers use racial categorization in collecting demographic data.
8. More precisely, to pass the naturalization test and become a naturalized citizen requires 5-year residence in the United States (3 years if the resident is married to an American citizen), a test of knowledge of the government and institutions of the United States, and the ability to understand English.
9. Prior to the McCarran–Walter Act of 1952 many nonwhites, especially Chinese and Japanese, were not able to become naturalized U.S. citizens.
10. An interesting illustration of just this issue is the voting in the First Supervisorial District in Los Angeles County, where approximately 80,000 votes were cast in a district with 1.9 million residents but many fewer registered voters, as the district has a large noncitizen population (Clark, 1998).
11. Despite the positive interpretation of electoral participation in this discussion, it is also clear that there is considerable political apathy among some immigrant groups. Some groups feel marginalized and that voting does not accom-

plish anything, while others report that they are overwhelmed by other responsibilities and do not have time to participate (de la Garza Menchaca, and DeSipio, 1994).

12. The latest naturalization figures list more than 600,000 new U.S. citizens in 2000. The number averaged about 230,000 a year in the 1980s, and in 1991 a total of 206,688 persons were naturalized. By 1994 a total of 543,353 immigrants were naturalized in a year (U.S. Immigration and Naturalization Service, 1999; see also J. Martin, 1995).
13. I would like to thank Karen Woodrow-Lafield for observations on naturalization rates.
14. This is according to the Current Population Survey (CPS) Combined Sample for 1998–2000. These results and those in this section on naturalization use a combined file from the last 3 years of the 20th century: 1998, 1999, and 2000. Because the CPS is a limited sample of approximately 60,000 households, I have increased the sample size by merging the data from 3 years. As the CPS retains households for 2 years and adds new households on a rotating basis, it is possible to retain nonoverlapping samples for 1998 and 1999 and the full sample for 2000, and so increase the sample size.
15. Much greater detail on the classification of the middle class is undertaken in Chapter 6. For the present chapter, I use the middle-class income definition based on 200–500% of the poverty level. This is close to the middle-income definition of Levy (1998).
16. The debates about whether previous political systems impact current behavior and whether they enhance or hinder current behavior are ongoing.
17. Parity scores are the ratio of Hispanic elected officials as a proportion of all elected officials as a function of Hispanic adult citizens as a proportion of all adult citizens.
18. However, both Skerry (1993) and Thernstrom (1987) argue that the evidence of Mexican electoralism before the extension of the Voting Rights Act is an indication that it was not needed.
19. See the U.S. Supreme Court Decision, *Easley v. Cromartie* (No. 99-1864) April 18, 2001, for a discussion of when cartographic manipulation may be allowed in redistricting.

CHAPTER 7

★ ★ ★

Joining a Divided Society?

A positive story of immigrant success emerges from the presentation of data on immigrants in professional occupations who are joining the ranks of citizens and participating politically—all signifying their movement into the middle class. It is much more than a story of the glass half full when we consider that many immigrants have been able to join the middle class even when the middle class as a whole is under financial pressure. Although not every immigrant will join the middle class, the numbers who do so are impressive and a testimony to the old adage that immigrants are doing what they always did, working hard and trying to "make it in America."[1] Immigrant progress is a strong belief within the foreign-born community, and the many media anecdotes are simply the stories behind the statistics.

At the same time we know that the current world in which contemporary immigrants move differs from the world in the era when earlier waves of immigrants undertook their long trips from Europe to the United States. Yet I am not sure that we should make too much of the differences; they may be more relative than absolute, and more in our minds than in the minds of the immigrants themselves. It is true that there have been economic and social changes in American society, especially in the large cities where immigrants still arrive and congregate in large numbers. Yes, the world is more interconnected than it was in the early 1920s, when the last very large waves of immigrants were arriving, and this may mean that it is easier to get from one country to another, but beyond that the main issue for new arrivals is still that of getting a foothold in the new society.

Getting a foothold in U.S. society is affected by the state of the economy in the period in which immigrants arrive and by the changing nature of income distributions in the United States. This chapter briefly examines the nature of income inequality for the population as a whole and how immigrants are doing within the changing income distribution.

"Globalization" is the common shorthand term describing the way in which the U.S. and world economy is being shaped by new production systems. It is also the shorthand designation for the new forms of global economic organization and the attendant processes differentiating markedly between workers endowed with diverse skills and education levels. The importance of human capital, and thus of education, is sharpening distinctions between the "have-mores" and the "have-lesses" in this regard. In addition, the late 20th century saw a move toward a reliance on market forces to respond to problems of underinvestment, both in human capital and in social and institutional innovation (Reich, 1991; Clarke and Gaile, 1998). This shift in economic policy was—and is—a move that places reliance back on individuals and local and private infrastructures, rather than on government, and in addition has removed much of the safety net that was available to previous aspirants to the middle class. It is worth reminding ourselves that the historic expansion of the middle class in the 1950s in the United States occurred in a time of expanding government (Levy, 1998).

Two particular aspects of U.S. and world economic changes are important for the discussion of current and potential immigrant success:

First, contemporary immigration is occurring in a world in which manufacturing jobs have been largely replaced by jobs in the information society. The relatively stable, often unionized jobs in manufacturing that were common 50 years ago have been replaced either by high-technology development jobs or by very low-skilled service jobs. Overall, the U.S. economy has lost several million jobs in manufacturing in the three decades between 1960 and 1990, just the period of recent large-scale immigration (James, 1995). At the same time, the U.S. economy has generated a vast service sector of low-skilled and relatively low-paying jobs, which have little if any security and even less chance of occupational advancement. There are frequent references to the emergence of a dual economy in which many of the native born and some immigrants are doing well while many others are working as janitors, cooks, and maids (Waldinger and Bozorgmehr, 1996).

Second, during this recent period of intense immigration, large cities have experienced major contextual changes. The combination of national budget cuts and local impacts of global economic changes have destabilized many local economies (Clarke and Gaile, 1998). Others have gone so far as to argue that the U.S. national administration has abandoned the cities (Rich, 1993). While American cities struggled to reorganize their tax bases and create new responses to global economic changes, some of the "gateway cities" in high-immigrant-impact states were faced with particularly trying situations—intensifying social polarization and segregation. In a sense, recent immigrants have been joining a more divided society in which the differences between the wealthy and the poor are especially influential in shaping future prospects of otherwise ambitious and energetic newcomers who have sought to improve their fortunes beyond their regions of birth.

There is by now a fairly substantial literature that suggests that, indeed, social polarization and associated spatial segregation in both developed and developing economies is increasing (Bourne, 1996; Reed, 2001). This seems to be especially true in the world's global cities, where there is the greatest separation between the rich and poor (Goldsmith and Blakely, 1992; Barclay, 1995; Fainstein, Gordon, and Harloe, 1992). The emergence of the hourglass economy in which there are increasing numbers of rich and poor (Sassen, 1994) and declining numbers of the middle class may be an overstatement (Hamnett, 1994; Clark and McNicholas, 1995). However, there is some support for a finding that the most wealthy individuals have gained considerable wealth in the economic changes of the 1990s. Use of the term "polarization" as a description of the outcomes of economic changes is going too far, but it is clear that the income distributions manifest greater inequality now than in the 1970s. The trends have intensified global competition, which in turn has increased unemployment levels and undercut real wage growth (Bourne, 1996). In addition, the impermanence of employment, apparent in part-time contractual and low-wage jobs, has played a role in changing the nature of income distributions, as have moves away from policies which attempted to redistribute income from higher to lower groups.

There is also some evidence that immigration itself is contributing to the increasing income inequality in the United States (Partridge, Levernier, and Rickman, 1996; Reed, 2001). Earlier research had suggested that the effects of immigrants were negligible (Butcher and Card, 1991), but a study by the National Research Council (Edmonston and

Smith, 1997) estimated that new immigrants reduced the wages of those without a high school education by about 5%. Other studies suggested that the effects for those with a high school education or less was closer to 10% (Camorota, 1997, 1998). There is also evidence that competition between the native born and immigrants in California was responsible for job losses, especially for those with less than a high school education. Moreover, the prolonged economic expansion in the United States in the 1990s created jobs mainly for that section of the U.S. labor force which has had at least some college (K. McCarthy and Vernez, 1997; Pigeon and Wray, 1998). Persons without a college education filled only 6% of the new jobs. Yes, new immigrants are getting a foothold in the new economy; it is their future trajectory in the labor market which is being debated.

Just how much of the inequality is being created by displacement and competition is not clear (Tienda, 1999). Recent research suggests that there is support for the idea both that "immigrants take low skilled jobs formerly held by natives" and that "immigrants also help push natives upward in the occupational stratification system" (Rosenfeld and Tienda, 1999, p. 97). On the one hand, the very high unemployment levels of native black workers in contrast with Mexican workers in the same occupations in Los Angeles is reasonable proof that displacement is occurring. On the other hand, it appears that there have been real gains by blacks in managerial roles in Los Angeles and Chicago. These outcomes would certainly feed into the possibility of increasing income inequality within ethnic and immigrant groups.

INEQUALITY AND ITS OUTCOMES

Within the debate about whether immigrants will make it to the middle class, there is a broader debate about whether middle-class status remains as accessible as before. It does seem that while the middle-class standard of living has become an icon in American culture, there appears to be a declining ability to attain it, certainly by a single breadwinner (Leigh, 1994). Earnings for the majority of U.S. households have not gained very much against inflation since the early 1970s (Newman, 1993; Levy, 1998). Yet, as the average wage has been declining, the very wealthy have been making unusually large economic gains. The top 1% of American households now account for about 37% of the private net worth of the United States. The percentage of households earning in the middle-income ranges has declined (Table 7.1), and the prospect for attaining home-

TABLE 7.1. Proportion of U.S. Households with Incomes in the Lowest, Middle, and Highest Quintiles

	1980	1990	2000
Lowest quintile (< 20%)	4.3	3.9	3.6
Middle quintiles (20–80%)	52.1	49.5	47.1
Upper quintile (> 80%)	43.7	46.6	49.4
Median income (Adj. $)	34,538	36,770	37,005

Source: Money Income in the United States, Current Population Reports, P60-200, 1998, and Current Population Survey, 2000.

ownership, the most important item of the middle-class dream, is increasingly dependent on both husband and wife working full-time (Myers, 1985). In households without two earners, the chances of entering the homeowner market are significantly lower. The outcome is such that Leigh (1994) suggests that the broad middle class is split apart into increasingly divided lower- and upper-middle-class segments. Here, the "haves" are workers who have an employed spouse; the "have-nots" are those who are single breadwinners.

Most recent studies reveal a decline in the size of the middle class. Levy (1998) used a general income category of $30,000 to $80,000 for ages 25–54 to capture prime earning years and showed that between 1973 and 1996 the proportion of white non-Hispanic middle-income families decreased from 66 to 55%. At the same time, both the number of people in poverty and the number of people at the upper-income levels increased. Overall, Levy finds that the income distribution is more spread out, the middle class is being "squeezed," and there is an increasing gap between rich and poor.

Both anecdotal and U.S. Census data on incomes suggest that the upward trend in incomes and wealth may be far less certain than it was just two decades ago. The ingrained view that new entrants to the labor market would improve on their parents' socioeconomic position may no longer be true. Newman (1993) describes the anxiety of the baby boomers, who are having more difficulty moving up the economic ladder, and in turn are now concerned about whether they will be able to provide more for their children than their parents' did for them. There is concern that the children of the baby boomers will have to settle for less—for poorer-quality schools and residence in less-affluent communities. Some baby boomers with good jobs and often two incomes are worrying about whether they can support the lifestyle they grew up with.[2]

Overall, the consumption package that we associate with the middle class is becoming less affordable. Housing, health care, and education are increasingly expensive (private school costs are now a significant part of the middle-class lifestyle in metropolitan areas, where the public schools have declined in prestige and quality). A major part of the increased cost of the middle-class package is the cost of housing. Younger households are having difficulty in entering the owner market. There has been an approximately 7–8% drop in the ownership rate. Married couples entering the housing market in the 1990s are older, on average, than those who entered the market three decades ago, a further indication that it takes longer to achieve the financial security which makes homeownership feasible.

To reiterate, it is in the context of a squeeze on middle incomes that the United States is absorbing the largest immigrant influx since the early years of the 20th century. The country is now engaged in incorporating the newest and largest wave of immigrants since that time. But that incorporation will occur in a very different social milieu than even three decades ago. While the American myth of rising with merit from humble beginnings as a result of individual hard work is still a core part of the American ethos (it remains the American Dream), the emphasis on individualism and the move to less government has created a changed political and social climate.[3] The rise in inequality, reduced welfare benefits, and fewer social support services will undoubtedly influence the process of assimilation, but in just what way is not at all clear as yet.

Thus, the process of integration is occurring as changes in household and family incomes occur, and against the backdrop of changes in the economy. While median household income has increased only slightly in the past three decades, there has been considerable variation from one period to the next (Figure 7.1). After the recession of the early 1980s, median incomes increased steadily from 1983 to 1989, and then declined during the recession in the early 1990s to nearly the same as the 1980 medians. The economic recovery after 1993 has brought median incomes back to levels slightly above those of 1990. The fact that a very large number of new immigrants have entered the United States to stay during this 20-year period has subjected them to a roller coaster of income increases and decreases, with very little progressive change for those who entered two decades ago. At the same time, those who entered the United States in the early 1990s entered at the beginning of a decade-long economic expansion, with all the associated positive economic outcomes of an expanding economy.

Median household incomes in the United States vary by region, fam-

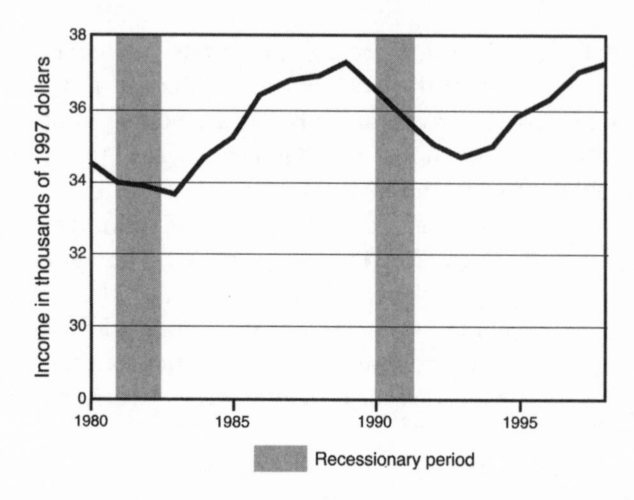

FIGURE 7.1. Median U.S. household income, 1980–1999. *Source:* U.S. Bureau of the Census, *Money Income in the United States,* P60–200, 1998, and Current Population Reports, 1998.

ily types, and the age of household heads. Medians vary from less than $20,000 to more than $50,000, depending on the region and the ethnic or immigrant group (Figure 7.2). Median incomes are higher in California for the total population but larger in New York for four-person families, and there is nearly an $8,000 difference across the states in median incomes. Median income is higher for older age groups, an unexceptional finding, and lower for female-headed households and single women. The range for immigrant groups is equally large, from a little more than $20,000 to nearly $50,000. Native-born Asians have the highest household median incomes, and foreign-born Hispanics the lowest. Thus, how groups will do in the future is affected by where they are residing and their household composition, in addition to their human capital.

PROGRESS AND INEQUALITY

There are a number of indices, ranging from simple ratios to the Gini coefficient that can be used to assess how the distribution of incomes is changing.[4] In essence the measures are designed to provide an index of whether the distribution of income across the population is becoming more or less even over time. Although the Gini coefficient is perhaps the most commonly used, it requires large sample sizes and is useful at na-

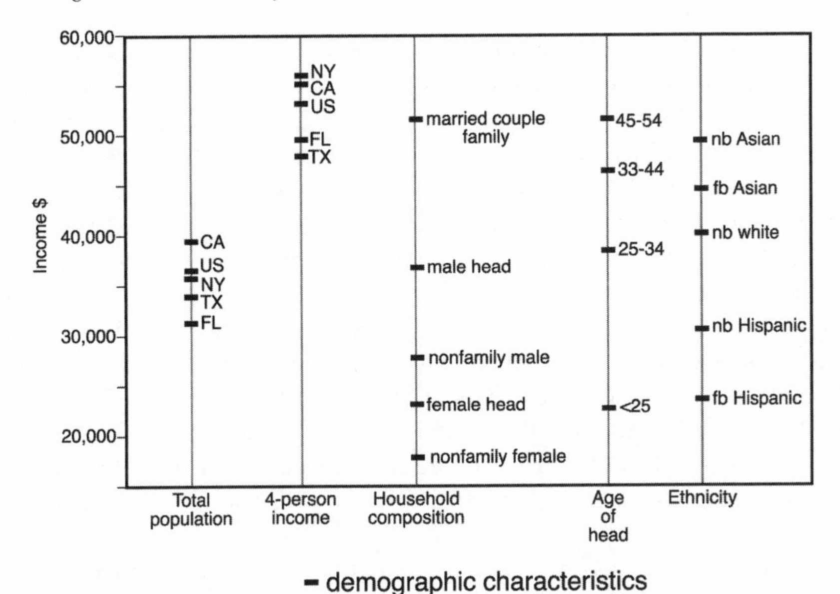

FIGURE 7.2. Household income variation by selected states and for age and household composition for the United Sates in 1997 (nb, native-born; fb, foreign-born). *Source:* U.S. Bureau of the Census, *Money Income in the United States,* P60–200, 1998, and Current Population Reports, 1998.

tional and regional but not local scales. In contrast, a simple ratio of inequality can be constructed by computing a ratio of the lowest to the highest income quartile for households. This ratio, say, the 90th to the 10th percentile, or (in the analysis used here) the 75th (top quartile) to the 25th (bottom quartile), is a good way of simply measuring the levels of inequality in income distributions, and also makes it possible to consider the way in which inequality has changed over time. The ratio of the 75th to the 25th percentile of income asks whether the bottom and top segments of society are about equal in their relative positions. Income at the 25th percentile captures how the lower-middle class is doing. The 75th percentile measures how the upper-income cohort is doing.

We know that income is not distributed evenly and we expect the ratio of the 75th quartile to the 25th quartile to be above 1, as there are more lower-income earners than high-income earners. The importance of the measure is that it can be used to track changes in inequality over time and to compare one group with another. If the ratio decreases, income is more evenly distributed; if it increases, income is less evenly distributed.

Previous studies of income distributions in the United States and California used the quartile ratio to show that income inequality has been increasing at both the national level and for California (Reed, 1999). Inequality took a fairly big jump after 1980 in the United States as a whole and earlier than that in California (Figure 7.3). Upper-middle-income households (the 75th percentile) in the United States in the late 1960s had about 2.3 times the income of the lower-middle class—the ratio of the 75th percentile to the 25th percentile. Households in California were at about the same point. The difference between the rest of the nation and California was small and remained that way into the middle to late 1970s (albeit with a small peak about the time of the oil crisis in 1973/74). After that time the inequality ratio diverged rapidly, reaching a ratio of 3.6 in California before settling back to about 3.4. In contrast the ratio for the United States as a whole moved up more slowly and is currently about 2.8.

Studies of increasing income inequality suggest that much of the increase in inequality in states like California—those with large-scale immigration—can be attributed to the new immigrants. An analysis of the changing distribution of male annual earnings shows that recent immigration accounts for a significant part of the variation in inequality levels across U.S. Census regions (Reed, 2001). By measuring the direct effect

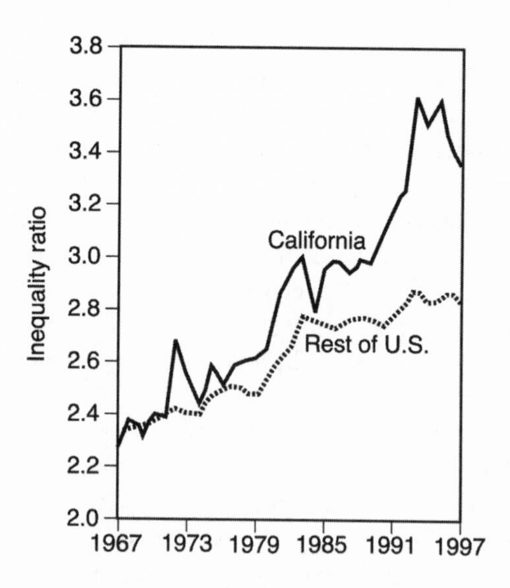

FIGURE 7.3. Inequality in U.S. and California incomes, defined as the ratio of the 75th percentile to the 25th percentile. *Source:* Reed, 1999.

of the presence of immigrant workers the research suggests that 13% of the growth in the size of the Gini coefficient, that is, increasing inequality, can be related to the presence of recent immigrants. The story is consistent with research showing that newer immigrants are doing less well, on average, than earlier arrivals and have lower wages. These findings are the setting in which we can examine how immigrant households are doing over time. We expect inequality to increase over time, but by how much in the absolute, and for which immigrant households, in comparison with the total population of households?

Using the percentile ratio both for the total population and for subgroups of the foreign born provides us with a rich set of results about the levels and changes in income inequality.[5] As expected, inequality ratios have increased, for all households, for the foreign born in the United States and across the four high-immigrant-impact regions. The largest increase and the greatest inequality are in New York/New Jersey, followed by Texas/Arizona/New Mexico. Apart from New York/New Jersey, the increases are modest for all households. The statistics for the foreign born are much more variable (Table 7.2).

Inequality ratios are higher for the foreign born than for all households but have increased only at about the same rate as all households. The inequality ratios are especially high in New York/New Jersey (and they increased significantly), and for most other regions they are only about a third of a point higher than for all U.S. households. The most interesting results occur when the foreign-born population is disaggregated by origin. For the nation as a whole, Asian foreign born and white foreign born are nearly stable in their inequality ratios. However, the patterns are complex on a state-by-state basis. Without considering both geography and ethnic origin it is impossible to tell the whole story.

In California, inequality has increased for both Asian and Hispanic foreign born, but the white foreign born have had less inequality in the two-decade period. In New York/New Jersey all ethnic groups have experienced significantly increased inequality. The most well-off foreign-born immigrants are increasingly different from the lowest earners. In Texas/Arizona/New Mexico both Hispanic and white foreign-born households have significantly less income inequality. The very large increase for Asians in Texas may be the result of small sample sizes, however. Hispanics in Florida have had decreased income inequality over time. The fact that inequality is not as great in 2000 as it was in 1980 can be broadly interpreted as evidence of both lessening inequality and upward progress of all immigrants. If only the wealthy were doing well, there would be greater inequality. Of course, it is also possible that those

TABLE 7.2. Measures of Income Inequality for Households, 1980–2000

	1980	2000
United States		
All households	3.11	3.37
Foreign born	3.57	3.75
Asian	3.22	3.28
Hispanic	3.10	3.28
White	3.82	3.80
California		
All households	3.08	3.40
Foreign born	3.21	3.49
Asian	3.24	3.32
Hispanic	2.75	2.92
White	3.71	3.55
New York/New Jersey		
All households	3.21	3.90
Foreign born	3.58	4.22
Asian	3.00	3.72
Hispanic	3.19	3.84
White	3.92	4.19
Texas/Arizona/New Mexico		
All households	3.13	3.43
Foreign born	3.49	3.29
Asian	3.43	5.31
Hispanic	3.35	2.88
White	3.61	3.21
Florida		
All households	3.07	3.25
Foreign born	3.50	3.50
Asian	3.33	3.47
Hispanic	3.42	3.39
White	3.33	3.79

Note. Inequality is measured as the ratio of the 75th percentile to the 25th percentile of household income. *Source:* U.S. Census of Population, Public Use Microdata Sample, 1980, and Current Population Survey, 2000.

with higher incomes are not doing as well, but the evidence from the income gains given in Chapter 3 negates this observation. The cause must be an overall gain in incomes. However, the serious downturn in the U.S. economy in 2001–2002 may well further impact income inequality.

These results temper the concerns about a bifurcating immigrant population. It is true that the absolute ratios are higher than for the general population, but this is an outcome of the overrepresentation of immigrants in the bottom and lower middle of the income distribution (Reed, 1999, 2001). At the same time, the fact that the ratios in some regions have been relatively stable, or even decreased, suggests that the foreign born are not any more affected by internal inequality than is the population as a whole. It is true that in New York/New Jersey, where large numbers of low-skilled Hispanic immigrants are struggling in poverty, the inequality ratio has seen its greatest increase. In Florida and Texas/Arizona/New Mexico, in contrast, the inequality ratios have actually decreased for Hispanics. The overall inequality ratio has increased only modestly. Clearly, immigrants as a whole are doing well in some locales, and some groups are doing better than others. The story is complex and nuanced, and it is much too easy to focus on only one part of the narrative and to ignore other equally important parts.

Income distributions provide additional understanding of why inequality might have increased in some contexts. Income distributions for the total population have become more peaked and shifted closer to the left axis. This is also true for the foreign-born population (Figure 7.4). But the data show that the biggest income distribution change is for Hispanic households. The shaded area in Figure 7.4 captures the increase, and the curve itself documents the rise in low-income households (the shift to the left in the graph) that is occurring in the foreign-born Hispanic population. There has been a significant increase in the proportion of low-income households, and when this is played out in particular locations, such as New York, there are dramatic impacts on income inequality.

The paths for the future will be a function, in large part, of the current and continuing immigrant inflows. If large numbers of less-skilled and less-educated Hispanics and Hispanic households arrive in the next decade, and if it is difficult for them to follow the optimistic paths we have examined in the previous chapters, there will be an increase in inequality. The rising inequality, not just for the population as a whole but for Hispanic households, too, hints at the bifurcation which is already a concern of the society at large. If the divide within the ethnic community also widens, we may well be following the divergent paths of progress

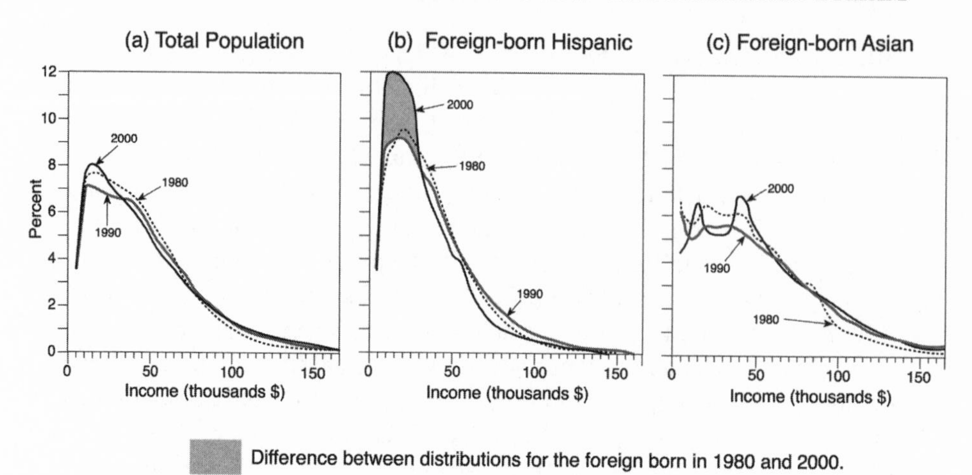

FIGURE 7.4. Income distributions by immigrant status in the United States. *Source:* U.S. Bureau of the Census, Public Use Microdata Sample, 1980 and 1990, and Current Population Survey, 2000.

and stagnation. Without considerably greater commitments to basic education and to innovative education programs for both younger and older immigrant households, it is possible that the immigrant community, like the native-born community, may follow a path to two nations, unequal and increasingly separated.

CONCLUSIONS

The evidence in this chapter further buttresses our general thesis of immigrant progress. There is no denying that there is a tendency to income polarization and to greater numbers of wealthy and poor households. But even in the tension of this change in the income distribution, the data in this chapter suggests that, on balance, the foreign-born population is at least holding their own within these societal changes.

Of course, the study does not argue that every immigrant household is doing well and that all immigrants are bound for middle-class status. Clearly this is not the case, and increasingly low-skilled, poorly educated households will have a difficult time making the progress that many in these cohorts had previously made. But in the context of the changes for native-born households, many immigrant households are doing extremely well. They will form the basis of a changed middle class, and by extension the stability that has been a central part of the economic advance of the society as a whole.

NOTES

1. There are also locality-based studies which make the same argument (Garvey, 1997).
2. We also know that expectations have risen, which further confounds the problem of measuring changes in the middle class.
3. It is clear that the creation of the large and successful middle class was not just a function of individual hard work and an advantageous economic climate. Government investment during the post–World War II period in the form of low-interest mortgages enlarged the homeowning opportunities, and the investment in the transportation infrastructure enlarged the urban areas available for suburban development. These programs were designed to stimulate the housing industry and in turn helped create a substantial middle class. Federal involvement in the creation of the middle class of the 1950s and 1960s was every bit as much an intrusion into the natural dynamics of the market as any poverty initiatives taken since that time have been (Newman, 1993).
4. The Lorenz curve and the Gini coefficient are the conventional measures used to calculate the difference between an actual distribution and a perfectly equitable distribution. A 45-degree diagonal line describes a population in which each portion of the population earns an equal portion of the income. The actual distribution, varying from that line, is the Lorenz curve. The Gini coefficient is a measure of the amount of deviation.
5. Income inequality is measured for households, and there is some debate about the proper unit to use for measuring changes in inequality. Most Census Bureau measurements of poverty and income quintiles and quartiles are based on household units. It is true that there are not the same numbers of people as there are numbers of household units in each quartile. The lowest quartile has less than one-quarter of all the population. Many low-income households are small—often single persons. In contrast, higher-income households often have more than one income earner. There is also an imbalance in the number of earners in each group. The lowest quartile has fewer persons of working age. Thus, in some senses the income gap, or the inequality, may not be as great as suggested by using a household analysis. However, to the extent that much of the analysis here is concerned with relative change, the contrast between individuals versus households should be seen as a caveat rather than a finding that negates the general results.

CHAPTER 8

★ ★ ★

Reinventing the Middle Class

Paths to the Future

The earlier chapters on professionalization, entry to homeownership, political participation, and middle-class achievement recount all that is positive about immigration—for the immigrants and for the United States. Some of the foreign born are working their way up the economic ladder, and others, starting out further up the ladder, have made even more progress. Their children are now participating in the opportunities that the United States has to offer children of successful families. Of course, many immigrants have not reached the middle class, but the evidence of the progress of specific age cohorts is that many are making real progress in the quest for their American Dream. But what is the evidence that the process can continue? What are the barriers that might stand in the way of continued progress for the foreign born, who continue to arrive in large numbers in the gateway cities of the large-immigrant-impact states?

Clearly, money matters and, just as clearly, the path to increased income is the ability to translate more education into professional or other higher-paying jobs. It is the gains in income and assets which bring a comfortable middle-class lifestyle. These gains are important because they enable the immigrants to buy housing in communities which are outside of the central city, away from problems of crime, poor-quality schools, and other negative societal externalities. In short, money enables immigrants to escape poverty neighborhoods and make the shift to middle-class neighborhoods, with better schools, more caring communities, and higher-quality housing.

A child's success is related to the success of the parents (G. Duncan and Brooks-Gunn, 1997). While the correlation is not perfect, there is a strong link between parental success and the outcomes for their children. We know, for example, that growing up in a poor family has a negative effect on later success in the labor force (Haveman and Wolfe, 1995). Earnings are lower, and the growth in earnings is slower. Thus, the continued success of the foreign-born middle class, and the ability of the children to enter the middle class, to reinvent the middle class, if you will, is bound up with the success of the parents. The achievements of the current foreign born will determine the probabilities of success for their children. To reinvent the middle class with the new foreign-born population requires a recognition of the context within which immigrants can make it to the middle class. For several decades, the middle class had help, tax advantages, good public schools, and reasonable access to health care. For the foreign born, some of these advantages are now in jeopardy, and a renewed commitment, especially to education, is critical for future success. Everyone recognizes this necessity, but few constructive policies are yet in place.

Thus far, the creation of a middle-class society has been one of the remarkable outcomes of the economic expansion of the United States after World War II. Even in a period of reduced or stable incomes, the United States has somewhere between 33 and 40% of the population in the middle-income range. It is true that expectations have increased and that many of the current middle class feel "relatively" less wealthy, but in essence there has never been such a large number with access to a lifestyle which would have been for only the privileged a half century ago. In this context, it is useful to recall that a very large proportion of the world's population lives on less than $2.00 (U.S.) a day and that the wage in rural Mexico is less than $1.00 per hour. In this perspective, and even in the face of very large local costs of living, a U.S. median income of nearly $40,000 is a very significant income indeed. It is little wonder that the pull of middle-class society, from inside and outside, is so strong.

Even the working poor, by U.S. government standards, often live well. To give the statistics a more immediate and real-world interpretation, it is useful to consider the case of Susanna, a single mother living in Southern California, whose income is below the official poverty line of

$17,000. However, she owns a small home, a 5-year-old car, a TV set, a video cassette recorder (VCR), and a computer. Even more important, her two children have health insurance and go to reasonable-quality local schools. It is true that Susanna's divorce settlement allowed her to keep the equity in the house and continue the mortgage payments, and that she has some child support. Even so, hers is a remarkable story of the opportunities available, even in high-cost housing markets in large cities of the United States in the late-20th century and early-21st century. Susanna may not be middle class, clearly not under the definition used here, but the issue is the future status of Susanna's children. If they can continue to attend good schools, graduate from high school, and attend college, then they are poised to continue the progress to the middle class.

The issue for Susanna and for countless households like hers, as well as for others that are more standard two-worker households, is the barriers and constraints that may limit her and her children's opportunities in the future. Will the society continue to provide reasonable-quality education at a reasonable cost? What are the impacts of continuing flows of low-skilled immigrants on job opportunities? Will local municipal economies be able to provide the social and other support networks for their populations? These are questions about the infrastructure that may in the long run be the most critical dimension of future immigrants' success.

PEOPLE AND PLACES

Individuals, families, and households arrive in the United States and become middle class in a wide variety of ways: many have education and skills before they arrive; others gain them by hard work after they have come to the country. Farah and Jay are nearly perfect examples of the processes that are pivotal to the growing immigrant middle class; Juan and Rosa exhibit contrasting but no less important elements of the middle-class story. Both are stories from Southern California, but they can be replicated in Dallas, Miami, and New York. They are both individual stories and examplars at the same time.

People

Farah and Jay and their two children arrived from Sweden in 1995. They were refugees from Iran in 1984 and lived in Sweden for 11 years before immigrating to the United States.[1] Now in their 40s, they both attended college in Iran. Farah completed a year of college, and Jay had undergraduate and professional training in accounting. They both undertook more education in Sweden (apart from learning Swedish), Farah in cosmetology, a professional training beyond traditional hairdressing, and Jay in accounting. The children, a daughter born in Iran and another girl born in Sweden, attend California State University and Beverly Hills High School, respectively.

The links to Southern California were through Farah's elder brother, who fled the revolution in Iran in 1979 and moved through Turkey and Italy to Southern California. It was those links that eventually provided the basis for the couple's "green cards" and immigration to the United States. Again we see the role of immigration links and channels, and the enabling process of earlier pioneering immigrants, as is true for many of the Iranian residents of West Los Angeles. Links are important, but the foundation of the couple's socioeconomic success was laid in Iran, where the schools required English as a second language, and their professional business and college training in both Iran and Sweden. Links and human capital are the crux of future success.

Farah entered the workforce as a contract hairdresser working for a large chain. Now she operates her own small salon, working 6 days a week from early morning till early evening, and later during holiday periods. In contrast, Jay must still pass the Certified Public Accountant (CPA) exam to become fully licensed in California. His professional training is equivalent or beyond that of his coworkers, but as with so many immigrants it is not straightforward to translate his skills from one society to another; when he does attain professional certification, the income gains will move the family to upper-middle-income status or above.

Currently, Farah and Jay do not fulfill our strict definition of middle class, as they rent rather than own a home. Acquiring the funds for the down payment and searching for a house take time. At this point, they rent a home in Beverly Hills, where there is an excellent school system, perhaps one of the best public school systems in the Los Angeles metropolitan region, and where there is a thriving Iranian community, which has established a very strong presence in West Los Angeles. The couple's attention to schools and education is reflected in their decision to choose the Beverly Hills residential area for its schools so as to ensure the superior education of their daughters.

Farah and Jay are not yet U.S. citizens but will eventually need to face the problem of dealing with dual citizenship, for themselves and for their daughters. Family connections—sisters in Sweden and Iran—

further emphasize the growing global community of immigrants and their strong linkages beyond a single country.

Juan and Rosa are more typical of the foreign born who continue to be the single largest immigrant group arriving in the United States—Latinos from Mexico and Central America. Juan came to the United States as a young adult with his sister from the Mexican state of Sinaloa in 1981. Both had received a basic high school education in Mexico. (Recall that a grade school education is typical for Mexican immigrants.) Juan's initial employment experiences ranged from construction work to gardening services. His reasonable ability in English, his willingness to work very long hours, and his entrepreneurial spirit led to the other path to middle-class status—that of self-employment. Even so, it is a fluid status: organizing and overseeing projects from simple gardening to sprinkler installations are complemented by working for other coethnics on projects which require more skills than Juan currently possesses.

Important additional elements of the story of an emerging middle-class Hispanic population are embedded in Juan's decisions and involvement with the larger ethnic community. His ability to transform his skills into self-employment was enhanced by his sister's ability and willingness to do simple accounts. As Juan points out, running even a simple gardening business requires basic managerial organization and simple record keeping including bookkeeping, skills which the extended family were able to provide.

On what is still a lower-middle income, Juan and his sister bought a home with their father's help nearly a decade ago, after being in Southern California for 10 years. It is a modest house in the San Fernando Valley, one of the tract homes built in the late 1950s for the burgeoning suburban white population that was moving out of the center of Los Angeles. Now that white population has been replaced by a mix of Hispanic and Asian households. (In the 2000 U.S. Census the San Fernando Valley was 42% Hispanic.)

An important dimension of Juan's successful advance to middle-class status was a delay in marriage. Juan waited until he was in his mid-30s before marrying. Following a classic pattern he returned to his home state of Sineloa in Mexico to find a bride, 24-year-old Rosa. Juan and Rosa do not yet have children, and they plan to have two, perhaps three. By delaying marriage and limiting their family size, they are fully aware of the greater chances for success they will offer their children. For them they see the possibility of more education and professional careers, a path that is increasingly the aim of foreign-born households in the United States.

These stories, and countless others of immigrant households across the United States (e.g., about young foreign-born academics), provide the human faces behind my statistical presentation. They do not replace the important numerical findings of how many new middle-class immigrants there are, and who and where they are, but they provide a sense of the people behind the numbers in two generally typical cases. Of course, the stories vary from place to place, and geography is important because it is the local context that sometimes enables and sometimes hinders the immigrants' progress.

Places

The opportunities and constraints are not uniform across the United States, and my state-by-state analysis has shown how much geographic variability there is in the inputs and outcomes. Because the cost of living varies significantly, it may be that local variations will have important effects on just who will gain and how the economic benefits will be apportioned.

From Oak Glen in Houston, Texas, to Glendale in Southern California, there are clear indications of the intersection of economic advantage in particular places, economic gains for both immigrant households and the communities they live in. The growing immigrant population in Houston has increased diversification in the economy by creating and growing small businesses, which in turn lead to movements up the economic ladder. Shopping centers, factories, and small businesses, owned and managed by new immigrants, are a central part of the transformation of the economy. People come, get an education, or increase their education, and many start small businesses. They want to settle down in suburbia and become Americans (Kotkin, 2002). The vibrant Armenian community in Glendale is made up of older Armenian immigrants who moved from East Hollywood and newer middle- and upper-class Armenians who arrived from Iran after the 1979 revolution. Glendale is perhaps one of the most visible symbols of Armenian immigrant upward socioeconomic mobility (Allen and Turner, 1997).

Still, there are measurable contrasts between thriving immigrant areas like Oak Glen in Houston, struggling immigrant areas like Weed Patch in the San Joaquin Valley near Bakersfield in Central California, and successful Indian entrepreneurs and computer scientists in Silicon Valley in Northern California. The children of Mexican immigrant meatpackers in Omaha, Nebraska, do not have the same advantages as children of Cuban businessmen in suburban Dade County, Florida (Miami). These variations highlight the constraints and opportunities for immigrants and their children. Another set of communities could be selected for Somali, Korean, Filipino, or several dozen other immigrant groups. While some groups in some locations are doing well, others are struggling. While some new immigrants with extensive education from their native countries are making scientific contributions to the United States, others are worrying about the next meal and the next job. Why are there such differences, and what is the role of the locality in the progress of immigrants?

Geography has always mattered because it encapsulates the constraints and opportunities that places provide. Weed Patch is symbolic of the countless small towns in the agricultural Central Valley of California, where undocumented immigrants work in the fields of vegetables and orchards of almond and walnut trees. But Weed Patch is replicated in vineyards in Virginia, in the rolling plains outside Lubbock, Texas, and any of the countless communities that rely on cheap labor to harvest the crops which array the aisles of the supermarkets of North America. The contrasts between the children of agricultural workers growing up in the small towns of rural California, Texas, and Virginia with the children of engineers, doctors, teachers, and scientists growing up in Palo Alto (California), Dunwoody (Georgia), and Boston is dramatic. Even at the state level, the contrasts between the income, occupations, and outcomes for Filipinos in Los Angeles and Dominicans and Haitians in New York City is a contrast between the opportunities of well-educated middle-class immigrants and those who are struggling to make ends meet.

The success stories of immigrants who are making it should not blind us to the other stories of poor immigrant families whose progress is slowed or halted both by their meager educational backgrounds and by the limited opportunities in their communities. Until the geography of opportunity is more equal, there will likely be a gap within the immigrant population, as there is in the native-born population. At the same time, recall the very positive findings of generational gains. The data are still sparse, but second and third generations are gaining as they, in turn, form households. As the data in Chapter 3 showed, later generations are more likely to be in the middle class than their parents, and younger groups move quite rapidly into the middle class. At the same time, the greater gains for some groups, especially Asians, points to the important role of previous education in creating the paths to middle-class status—a finding that has been stressed throughout this book.

CONSTRAINTS AND BARRIERS

Who (and how many) will make the continuing progress to middle-class status is dependent on a number of external forces: the quality of education, labor market opportunities and competition, and particular effects in local housing and labor markets.

The Role of Education

In the past decade, education, and especially education in inner cities, has emerged as a critical dimension of the future success of the American economy and society. There is no question that the acquisition and use of human capital is a critical dimension of immigrant progress. At various points in this book I have emphasized the fact that the American economy is in a transition state, a transition in which higher skills and labor market flexibility are more and more essential for employment and critical for higher wages. Lower-skilled immigrants, those without good educational backgrounds, are going to have real difficulty in gaining employment and, even more important, in being able to move up the occupational ladder. It is true that service jobs, often temporary and without benefits, are available, but these are not jobs on which to build a future, nor are they jobs which will provide the secure income to raise educated families.

To reiterate, theory and empirical analysis tell us that children who grow up in poor and poorly educated households are disadvantaged throughout their lives. The evidence of studies of children who grow up even in low-income households is that they take a considerable time to overcome the deprivation of those early years (G. Duncan and Brooks-Gunn, 1997). In some cases, the data suggest that those children are never able to make up for the disadvantages created by the lack of early attention to basic learning skills. Studies of Head Start programs for Latinos provide convincing evidence of the real gains that come from early intervention in education (Currie and Thomas, 1997; Garces, Thomas, and Currie, 2002). In their study, Currie and Thomas showed that participation in the Head Start program closed about a quarter of the gap between Latino and non-Latino white children, and two-thirds of the gap in the likelihood that the students would have to repeat a grade. Such children need help, and need help early, but so too do students who have difficulty completing high school or who have arrived in the United States without completing a high school education. Young-adult education and adult education are equally important in creating an educated labor force and increasing the human capital of the population as a whole.

Although there is now evidence that the foreign born are branching out, moving to cities throughout the nation, the major concentrations of the foreign born are still in the large gateway cities of a handful of states, which still absorb the major flows of immigrants. Thus, the issue of educating the children of the newly arrived immigrants will fall to the public education systems in these entry-point cities, which will bear the burden of the educational process. This situation is even further complicated by the fact that the native-born white population is increasingly leaving the public education system for private schools, or has already moved away from inner-city public school systems to suburbs well outside the metropolitan areas. A decade ago public policy analysts raised questions about the ability of the current school systems to provide the education for the new immigrant populations, and those questions are still relevant today (Rolph, 1992; Stasz, Chiesa, and Shwabe, 1998).

As I noted in an earlier study of immigration to California (Clark, 1998), the reality is that the state is absorbing and educating a very large proportion of all new immigrants who come to the United States—and some would say not doing a very good job of it either. It is clear that many of the immigrants coming to the United States lack the prerequisite high school education that affords some prospect of meaningful employment. In addition, their high fertility is creating a population that must be educated if they are to take a full role in the future of 21st century American society.

The changes in the U.S. economy, where occupational boundaries are less clear and many jobs are now temporary, has created a need for a more flexible workforce. The "new economy" has also focused attention on the close relationship between education and economic health, and how, as a nation, the continuing economic gains in the United States will be closely linked to education. The concern about international competitiveness has led to a number of initiatives designed to rethink education and its role in the economy (Stasz et al., 1998). While there is debate about whether the education system is failing the nation as a whole, there are few who would disagree with the observation that there is a real need for improvements.[2] Nowhere is this need greater than in the inner-city schools in those metropolitan areas where most new immigrants are settling (Vernez, Krop, and Rydell, 1999).

The challenge is to educate the new and diverse immigrant population, some of which is lagging behind the native-born population. While money will not solve all the problems of inner-city education, the low level of funding for education in California, Texas, and Florida and the inadequate funding for education in Illinois limits the ability to reduce class size and experiment with alternate educational programs.[3] Although education is a cherished local prerogative, there are now a number of moves to provide 21st-century alternatives to the basic educational structures that have been in place for most of the 20th century, even alternatives to the public education system per se. The New American Schools initiative is one such private nonprofit organization that was launched to effect school reform and respond to the perception that U.S. schools are failing students, especially in inner-city and poverty settings (Berends, Chun, Schuyler, Starkly, and Briggs, 2002). Even so, these alternatives may not be sufficient to solve the increasingly intractable problems of educating a multilingual and inadequately prepared foreign-born population. Bankrupt inner-city school systems in Baltimore, Detroit, and Compton (Southern California) and fiscally stressed inner-city school systems in a dozen other states are only the most visible signs of public education in trouble and the failure of school systems to address the needs of the new immigrant population. Federal resources may be a critical component of changing the educational system in the United States. Only by creating human capital can the nation provide the foundation that is essential to continuing the emergence of the new immigrant middle class.

In the long run, it may be that educational institutions are more important than either labor markets or the social-welfare system for creating community well-being (Reitz, 1998). It is through investment in education that society creates increased human capital, greater social well-being, and the context in which innovation and new ideas can flourish. In the past half century that process has created and continues to create the massive gains in human capital of the native born in the United States. Now, nearly one-third of the U.S. population as a whole have some college education or a college degree. This is a considerable increase from the only 17% of the population with some college education 30 years ago (U.S. Bureau of the Census, 1972, 1993). Providing those opportunities for the new immigrant population is a critical dimension of recreating a new middle class in the 21st century.

The Labor Market and Competition for Jobs

Although there is contention about whether or not immigrants compete with the native born for jobs as earlier chapters have discussed, there is growing evidence that later arrivals are competing with earlier arrivals for a place on the economic ladder. Although much of this competition is at the low end of the skill market, this competition is apparently an important concern for the new immigrant populations because it affects the ability of those already in the country to move up and make the economic gains that are so necessary to their continued progress to the middle class.

The studies by the National Academy of Sciences (NAS; Edmonston and Smith, 1997) discussed earlier, and a wide range of other studies (e.g., see Card, 1999), have now documented that overall there are only marginal impacts of immigrants on the labor force earnings of the native born, on average. At the same time the research shows that unskilled immigrants are as likely to hold the same jobs as unskilled native-born workers. The distribution of occupations for native-born high school dropouts and Mexican immigrants (who usually do not have a high school education) are very similar (Camarota, 2001). As a result, we can expect that there is direct competition between these two groups. Both the aforementioned NAS study and one by Borjas, Freeman, and Katz (1997) concluded that immigration has an adverse impact on the wages of the native born who do not have a high school education. Thus, there are negative impacts, but to the extent that only a small percentage of the U.S. population does not have at least a high school education, the effects must be small in an overall economic sense.

A related issue is the nature of the demand for low-skilled labor. If the demand for low-skilled labor decreases, that is, if there is a shortfall of jobs for low-skilled workers, then there is likely to be downward pressure on their wages. The available evidence tells us that the wages of high school dropouts declined about 7% between 1989 and 1999 (Camarota, 2001). During the same period, the wages of those who completed high school rose about 9%. Clearly, there are significant gains to greater human capital, as would be expected. At the same time, the data suggest that to the extent that there is a constraint on the future of the immigrant population, it is a constraint on the most low-skilled and least-educated section of the immigrant population. It may even be that the continued flow of these immigrants may create serious problems for the previous immigrants themselves and for the communities to which they move. Even with the best intentions, it is not possible to move up without jobs and incomes that are sufficient for basic family support, and more than that, the assets to provide the support for education and the increases in human capital that more education brings.

Those immigrants who make the greatest gains are those who arrive in the United States with more than a basic education. Even then, it may not be straightforward to translate previous education and skill levels to new jobs and higher incomes. Doctors and teachers from foreign countries often have difficulty in making the transition to middle-class professional activities in the United States. When they give up jobs within their specialty because their degrees or college credits do not meet American standards, they are quite clearly making big sacrifices for the good of the next generation. The anecdotes are legion: the South Vietnamese general who wrapped snack cakes for Krispy Kreme and whose children are college graduates from the University of California; the Colombian mother, a respected scientist, who overstayed her immigrant visa to work in an anonymous factory and whose daughter is in college. These immigrants are focused not on their own short-term goals but on their children's long-term success.

Many professionals do make it. Recall that Chapter 4 documented the large-scale increase in foreign scientists and their contribution to the U.S. scientific community as well as the gains they are making for themselves and their families. However, for those who do not manage the transition to professional occupations in the United States, those former professionals are competing with low-skilled workers, which will in turn delay their chances to improve their outcomes.

POPULATION GROWTH AND THE FUTURE

The changes we have been witnessing in the immigrant flows and in their socioeconomic outcomes are fundamentally imbedded in the changes occurring in the foreign-born population as a whole. They are both the generator and the outcome: they are creating most of the population growth in the United States and are impacting the present and future of those immigrants and their offspring who are here and those yet to come.

The immigrant flows and their relatively high fertility are creating a rapidly growing U.S. population. Thus, continuing inflows cannot but impact the size and eventually the quality of life for all—citizens and noncitizens alike. The demand for visas, green cards, refugee entries, and special labor force programs shows no sign of slackening. The requests from potential immigrants in nearby countries—Mexico and the Central American nations—and further afield—from China and Southeast Asian nations—are fueled by their large and often impoverished populations, but also by those members of their intellectual elites who would simply like to do better. Although it is difficult to assess exactly the implications of this rapid U.S. population growth, and it is an issue that is fraught with political sensitivities, it is important to recognize also that absolute population size will have an impact. It will almost certainly be very important for how well the current immigrant populations will be able to do in the future society.

Current U.S. Census predictions suggest that the United States population will grow to about 400 million by the middle of the 21st century, well within the lifetimes of many children and even young adults alive today.[4] There are two strongly held perspectives on the issue of the continued growth of the U.S. population. On the one hand, Californians for Population Stability (CAPS) and Beck (1994) advocate a substantial reduction in immigration or even a complete halt to immigration, arguing that we are already over the numbers that the nation as a whole and particularly certain states can absorb. They point to impacts on the environment, strains on the infrastructure, and the likelihood of a declining quality of life for all residents, native born and immigrant alike. On the other hand, some economists, including Alan Greenspan, see immigration fueling the economy, providing new workers, and reinvigorating the entrepreneurial spirit of the United States (Chavez, 1996). Still others note that the new immigrants will be the additional workers whose tax dollars will rescue the ailing Social Security system (Torres-Gil, 1998).

Of course, it is not possible to decide who is right; only time will enable us to truly evaluate the effect of large numbers of new immigrants. However, it is worth noting the levels and kinds of growth that are occurring, and the implications for costs and infrastructural investment which will be needed as a result of that growth. To the extent that the growth continues at current levels, the economy will need to grow at or above recent levels. The current population projections predict quite substantial increases in the U.S. total population. As mentioned earlier with current immigration levels, and even reduced fertility, the total numbers by the end of the century could be between 400 and 550 million people, an increase from the 285 million people in the United States today. Moreover, the growth will be disproportionately in Florida and the West—including California—in those states which already have large numbers of immigrants.

The growth is being created by both large-scale immigration, on the order of 1 million a year, and births to the second generation of the earlier arrivals and the current foreign born. About half of all births are now occurring to the foreign-born population, which makes up about 11% of the total U.S. population. These youthful populations will need education, social services, health care, and all the concomitants we have come to expect of a comfortable way of life. The growing youthful immigrant and second-generation population is balanced by an aging native-born white population. Thus, there will be a somewhat bifurcated young and elderly population coexisting in the same communities.

The outcome of such population growth will be simultaneous demands for increased budgetary allocations for education and for elderly social-service programs. An optimistic view suggests that the hard work of these new immigrants will help create the new society in which the foreign born can continue to make the economic gains that earlier arrivals made. Their gains will provide the tax dollars to support the Social Security system, just as the tax dollars of the current baby boomers provided the revenues to support the current system. I am fully aware of the arguments of Borjas (1998) and others (Beck, 1994) that a continuing flow of low-skilled immigrants will use more social services and consume more tax dollars than they provide, and I share the concerns that those studies raise. But as fertility drops and if the economy continues to grow, there are cogent counter-arguments that the current immigration flows will enrich and enhance the U.S. economy as earlier waves did. The question at issue is about the rate and level of current flows, and whether there is room for changes in the immigration policy that presently exists. Certainly there should be an informed debate on current immigration policies, but from the perspective of this analysis there is a great deal of evidence that, to repeat an earlier comment, immigrants are still seizing the opportunity to create their own American Dream.

Whether the continuing growth of the U.S. population will be a constraint on the success of the foreign born will depend as much on who the new immigrants are as on the raw numbers of immigrants. If many new immigrants arrive with already substantial human capital, they can immediately contribute to the economic growth of the United States. If many low-skilled immigrants arrive, they will need resources to make the leap to fully participating new members of the postindustrial economy. There are strong arguments in favor of an immigration policy in which the United States fulfills a humanitarian commitment to immigration without causing serious disadvantage to our current foreign-born and native-born populations.

CONCLUDING OBSERVATIONS

There are now nearly 4 million foreign-born and ethnic native-born middle-class households in America, approximately 12.3% of all middle-class households. The media stories of a growing ethnic and foreign-born middle-class population are stories of a real change in the demographic structure of the middle class in the United States. These families are homeowners, they are increasingly members of the professions, and they are participating in the political process. Overall, nearly 50% of foreign-born households are homeowners and for some ethnic groups the proportion is much higher. In some states the foreign born make up more than 20% of the professional occupations. Those proportions will increase as the foreign born gain more education and on-the-job training. Indeed, the foreign born are remaking the middle class, a middle class that is more diverse than it has ever been in the past.

Immigration is a process, not an event, and a process that is notably self-selective. It gathers from a population of those who tend to be young, ambitious, and energetic, those inclined to seek their fortunes beyond their home countries. Immigrants will continue to leave areas of deprivation and try to enter places of opportunity. Moreover, it is very unlikely, given the connectedness of the global economy, that it will be possible to stop those who truly wish to migrate. They will come with documentation or, if necessary, without it. Given that assumption, what are the dimensions of a future immigration policy and how can that policy stimulate rather than hinder the progress of immigrants toward the middle class?

The United States is still exceptional in that its immigration policy is largely based on ad hoc rules for admittance, rather than a coherent strategy of deciding an overall basis for the numbers and backgrounds of immigrants, and the mix of refugees, asylum seekers, and economic immigrants. The suggestions by the U.S. Immigration Commission[5] for modest controls on the numbers of immigrants, greater attention to education for immigrants who are already here in the United States, and some focus on the impact of immigrants on native-born workers are all worthy of further study. They are suggestions which would likely increase the chances for more immigrants, already in the United States, to follow paths of upward mobility. Whatever the final outcome of a discussion of immigration, it is important to shift the discussion and debate away from the simple extreme responses of "close the borders" or "open the borders." These alternatives only serve to polarize an already contentious debate.

This book has painted a fairly positive picture of the slow progress of immigrants toward middle-class status. The entry into the middle class by the youngest cohorts of immigrants and the progress across generations are useful counters to those who would see immigration only as a problem. Still, the overall slow increase in the middle-class Hispanic foreign-born immigrant population is cause for concern and requires specific responses if the society is to avoid compounding problems of alienation, poverty, and even an underclass population (Clark, 1998). In addition, the very large differences in middle-class gains across regions are a reminder that outcomes vary geographically and that immigrant outcomes must be seen in their geographic and ethnic contexts.

The findings in this analysis of the changing nature of U.S. society and the progress of the foreign born are cause for both celebration and caution. Indeed, there are those who only celebrate the continuing flows of immigrants to the United States and those who only worry about those numbers. The celebrators must recognize that only with specific political attention can the immigrant flows continue to be translated into a new and integrated force within the United States. The worriers need to reflect on the gains and contributions of the burgeoning foreign-born population. Numerically, Asian and Hispanic middle-class households are a new force in the demography and economy of the United States. They represent clear gains in upward mobility for a wide range of Hispanic households, gains that are greatest in selected regions such as California and Texas. At the same time, there appears to be a slowing in the rate at which younger cohorts are able to become members of the middle class. While there is considerable upward mobility, it is not as great for some younger cohorts, or in the New York/New Jersey region. The Hispanic groups in New York/New Jersey are quite unlikely to move upward in the same way as Hispanics in Texas and Florida have done.

This chapter began with the notion that the influx of the foreign born, the new immigrants who arrived before 1980, and to a lesser extent the immigrants who have arrived more recently, will become the new immigrant middle class. Indeed, the findings of this book suggest a continuing process of immigrant middle-class gains that have occurred in the past, are occurring now, and will occur again. Yet to be determined are just what the household composition will be and how assimilated, in classic terms, that population will be. However, the evidence seems to be that immigrants are doing today what they have done in the past: working hard, trying to "make it" in America, and becoming engaged citizens and productive workers in their new society. As they intermarry, and as American society becomes more like the blend already personified by Tiger Woods, Maria O'Brian, and Jennifer Lopez, we may well forget that race and ethnicity were once given prominence over the blending which long has been—and will remain—the centerpiece of a reconstructed middle-class American society.

NOTES

1. The decision of Farah and Jay to migrate to Sweden first was dictated simply by the ease by which immigrants can claim refugee status in that country.
2. There is a very large literature on education and educational reform. A substantial number of these reports which discuss the issues of education in general and education for inner-city school children in particular can be found on the RAND website at *www.rand.org*.
3. See *www.ed-data.k12.ca.us* for data on comparative state funding of education.
4. The U.S. Bureau of the Census projects population under a variety of scenarios through the end of the 21st century. Currently, the bureau's middle-level projection is for 403 million persons in 2050; the lower bound of the projection is 313 million and the upper bound is 552 million. The estimated lower-bound projection with continuing international migration is 327 million.
5. The U.S. Immigration Commission was chaired by former Representative Barbara Jordan of Texas. After her untimely death in January 1996, however, the Commission's report was not brought to Congress for action; it was eventually filed at the end of 2000.

APPENDIX

★ ★ ★

Data and Data Sources

Several different data sources have been used to construct the tables, graphs, and maps used in this study. But beyond the research sources used specifically in this book, it is worthwhile including a brief review of the nature and limitations of data that are available for the statistical analysis of immigration and immigrant settlement patterns.

Before examining the data sources some definitions are worthwhile repeating. The terms "foreign-born population" and "immigration" are related but do not mean exactly the same thing. An international migrant is a person who changes his or her "usual place of residence" from one country to another (U.S. Bureau of the Census, Current Population Reports, Special Studies, P23-195, 1999). Immigration can refer to all migration to the United States or more narrowly to the legal migration of non-U.S. citizens (as noted earlier, the technical term is "aliens") to the United States; "immigrants," as defined by the U.S. Immigration and Naturalization Service (INS), are aliens admitted for lawful permanent residence in the United States. The term "foreign born" denotes all individuals born in a foreign country except those who had at least one parent who was a U.S. citizen. The foreign-born population includes all individuals residing in the United States whatever their legal status (U.S. Bureau of the Census, Current Population Reports, Special Studies, P23-195, 1999). In general, in this book I use the terms "immigrants" and "the foreign born" interchangeably, though the data used are always for the foreign-born population.

There are three main sources of large-scale data for the study of the foreign born: the decennial Census of the United States, the annual survey known as the Current Population Survey (CPS), and statistics from the INS. Although data are given on the foreign born in the Panel Study of Income Dynamics (by the Institute for Social Research, University of Michigan, Ann Arbor), the very small sample size precludes the use of these data for any in-depth analysis of immigration.

Individual research groups have collected specific studies of immigrants and immigrant flows, and some of these data are available from the Inter-University Consortium for Political and Social Research (ICPSR) at *www.icpsr.umich.edu*. A notable data set on Mexican flows to the United States, especially from rural areas of Mexico, can be accessed at the Mexican Migration Project (MMP) at *www.upenn.edu/mexmig*. As reported at that website, the MMP71 database is the result of an ongoing multidisciplinary study of Mexican migration to the United States. It contains data gathered since 1982 in surveys administered every year in Mexico and the United States. There are five primary data files, each providing a unique perspective of Mexican migrants, their families, and their experiences. The database contains an initial file with general demographic and other files on migratory information for each member of a surveyed household. Detailed labor histories for each head of household and each spouse are available in separate files.

DATA FROM THE DECENNIAL CENSUS

The U.S. Census, conducted every 10 years, is a rich source of data on immigrants and immigration. The most recent Census data for 2000 are being released in the period 2002–2006. The data are reported in several formats including full-count data from the "short form" Census questionnaire (enumeration data) for large and small areas, and sample data on individuals and households. The data of most relevance for the present study come from the Public Use Microdata Sample (PUMS) for 1980 and 1990. The 2000 PUMS data will be released in 2004.

The PUMS data include two separate files: the 1% sample and the 5% sample of the "long form" questionnaires collected by the U.S. Census. The PUMS 5% sample is a subsample of the Census sample of approximately 16% of all U.S. households in 1990. The PUMS 5% sample provided the data used in the analysis in this book. Unlike summary data where the basic unit is a Census tract, county, or state, in microdata the basic unit is an individual household and the housing unit in which that family lives. In effect, PUMS files make it possible to construct analyses of individuals and their characteristics. The data can be manipulated as if they had been collected in a special-purpose sample survey.

The PUMS files contain the full range of population and housing information collected in the Census. In the analyses in this book the PUMS files were used to construct data on middle-class households who were foreign born, on the characteristics of foreign-born professionals and homeowners, and on the U.S. citizenship status of immigrants. However, as in the data from the CPS (discussed below), analysis of small geographic areas may not be feasible because of the small number of cases. More detail on the PUMS files is contained in the PUMS Technical Documentation. It is available for both 1980 and 1990 from the U.S. Bureau of the Census at *www.census.gov/main/www/pums.html*.

DATA FROM THE CURRENT POPULATION SURVEY

The CPS is conducted each month in the United States. The March CPS includes the basic CPS and supplementary questions. The CPS is a sample of about 65,000 households and 130,000 persons. The survey's estimation procedure adjusts the weighted sample results to agree with independent estimates of the civilian population of the United States. Because the CPS estimates are based on a sample, they will likely differ from a complete enumeration and, for microgeographic areas and even for some states, the small number of sample cases make estimates less reliable. However, for broad comparisons the data from the CPS are reliable and provide an excellent source of up-to-date information on a wide variety of population characteristics. Specific details on CPS estimates of the foreign born are available in *Profile of the Foreign-Born Population in the United States* (U.S. Bureau of the Census, Current Population Reports, P23-206, 2000).

The basic CPS collects primarily labor force data on the civilian population. The March supplement for the CPS includes the basic CPS questions and questions about poverty status, income received in the previous calendar year, educational attainment, household and family characteristics, geographic mobility, and data on the foreign-born population. The data on year of arrival and immigrant ancestry make the CPS particularly useful for national-level analyses of the foreign-born population. In the March 2000 CPS the sample was increased to include additional Hispanic households. The data are available from the ICPSR in Ann Arbor, Michigan, at *www.icpsr.umich.edu*.

Data on voting and U.S. citizenship were drawn from the CPS, Voter Supplement File, November 2000. This file includes data on registration and voting in the November 2000 election; the file was released in June 2001. In addition to U.S. citizenship, registration, and voting data, the file includes standard CPS measures of education, occupation, income, and year of entry for the foreign-born population.

In some cases in the analysis in this book, two or more years of CPS data were combined to increase the sample size. It is possible to merge successive years of data. The nature of the CPS data precludes a simple aggregation, however. In each year, half the sample in any month reappears in the sample a year later, after which the cases are dropped from the sample. It is possible with the file structure to drop overlapping cases and, in this way, a 3-year merge effectively doubles the size of the sample. This approach was used in the present study for the analysis of second- and third-generation immigrant success.

DATA FROM THE IMMIGRATION AND NATURALIZATION SERVICE

Data on legal immigration and immigrants is collected by the INS through the immigrant visa form (OF-155, Immigrant Visa and Alien Registration) for the U.S. Department of State. Information on those who adjust their status after entering the United States is collected from the I-181 form. In essence, aliens wishing to become legal residents of the United States use one of two processes to become residents: (1) Aliens living abroad may apply for a visa at a consular office of the U.S. Department of State; once issued a visa, they may enter the United States and become legal immigrants at that time. (2) Aliens living in the United States, including temporary workers, refugees, and some undocumented immigrants, may file an application for adjustment of their status. Details from these records are reported annually in the *Statistical Yearbook* of the INS, which is available online at its website: *www.ins.usdoj.gov*. Specifically the current and some past yearbooks can be accessed at *www.ins.usdoj.gov/graphics/aboutins/statistics/ybpage.htm*.

The INS *Statistical Yearbook* has detailed information on numbers of legal immigrants admitted by year, country of origin, and class of admission status. Class of admission status refers to whether the immigrant came under family preference categories, as refugee/asylees, or some other status. For example, in the year 2000, immediate relatives of U.S. citizens made up 41% of all legal immigrants, up from 15% in 1990. *Statistical Yearbook* data include considerable geographic detail as well as information on national origin and admittance status. No individual data are available in the *Statistical Yearbook*; however, the INS microdata files include a valuable small set of variables from the administrative application for lawful permanent residence. In addition to basic immigrant personal characteristics, these variables include the country of origin, visa entry status, place of intended settlement, and year of entry. These data are distributed through the National Technical Information Service and through the ICPSR website at the University of Michigan, Ann Arbor: *www.icpsr.umich.edu*. The U.S. Bureau of the Census uses them for population estimates, and several other organizations have begun using them to examine settlement patterns.[1]

A specific data set from the INS examines the applications of undocumented immigrants for regularization. The Legalized Alien Processing System data files for 1989 and 1992 (LAPS-1 and LAPS-2) have nearly 1.5 million individual records for undocumented immigrants who applied for regular immigrant status after the Immigration Reform and Control Act of 1986 (IRCA). In addition to data on processing status, the files contain information on occupation, income, and a variety of other socioeconomic characteristics. A more detailed discussion of these files can be found in Powers et al. (1998).

THE TERMINOLOGY OF ETHNIC STATUS

The federal government has Census categories for both race and ethnicity. In past Censuses the U.S. Bureau of the Census used "black" and "Asian" as racial categories and "Hispanic" for people of Spanish ethnicity. The exact terminology has not remained completely constant over time, and the wording of Census questions about ethnicity have changed subtly over the years. It is important to be cognizant of these changes, although in general it is possible to make broad comparisons of groups over time. Even so, the use of the word "Latino" along with "Spanish" and "Hispanic" on the basic Census question on Hispanic identity probably increased reporting in California. As Foner (2000) emphasizes, "Hispanic" is a Census term, created by the Census Bureau as a way to designate and enumerate the Latin American population in the United States. Most Latin American immigrants prefer to be identified by their country of origin and use their ethnic background as self-designation rather than Hispanic. Although the statistical creation "Hispanic" may indeed become a social reality (Foner, 2000, p. 156) if the public eventually adopts the term for everyday use; the word "Latino" seems to be gaining in popularity in the media of late. (I have used both terms in this book, generally "Hispanic" when discussing Census data.)

It is important to note that Hispanics are included in the racial category "white" but are distinguished by ethnic background depending on their nation of origin. Now "African American" is an official equivalent of "black" and "Latino" is being used by the U.S. government as an alternative designation for "Hispanic."

For the first time in the 2000 Census, the questionnaires allowed people to check off more than one of the ethnic identities in the questionnaire. The change in approach was a direct attempt to confront the small but increasing number of Americans and their children who identify themselves as of mixed racial and/or ethnic heritage. The numbers are still small but will grow in the coming decades. In 2000, less than 5% of the Californian population checked off more than one race (Allen and Turner, 2001, 2002). As most of the analysis in this book is about the foreign born, this is less of an issue than it is for studies of the native born.

COUNTING THE FOREIGN BORN AND IMMIGRANTS

Most of the analysis in this book is focused on the foreign born, people who were born abroad and later immigrated to the United States, and sometimes in comparison with the native born. In general this categorization of native born and foreign born is straightforward. Persons born abroad of U.S. citizens are classified as native born. Persons from Puerto Rico have automatic entry to the United States and are not considered foreign born. When questions about time of arrival are asked, however, the data are not necessarily so simple. Ellis and Wright (1998) have shown that the questions asked in the CPS and the decennial Census are slightly different and may yield small differences in the results. The CPS asks of the foreign born, "When did you come to stay," whereas the decennial Census asks for "year of arrival." Although the differences in most cases are likely to be small, a number of immigrants who immigrated, then returned to their home country, and later returned to the United States will likely answer the two questions differently.

QUALITATIVE AND QUANTITATIVE DATA ON IMMIGRATION

Most of the material in this book is based on detailed analyses of U.S. Census and other statistical material. There are a number of individual stories derived from interviews with individual immigrant households and from newspaper interviews reported in national and local newspapers. Both quantitative and qualitative research studies in the social sciences have informed our understanding of immigration and immigrant outcomes. Individual stories put a face on what is often a summary report in the Census or other statistical sources. In-depth interviews with small samples of new immigrants can provide a perspective on how immigrants feel about their new country and how they are adapting. At the same time, it is only through large-scale data analysis that we can provide an assessment of the picture of immigrant successes at a national scale.

It is clear even from the limited number of individual immigrant portrayals in this book that immigrants advance in a multitude of different ways and by different paths. The stories of these paths enrich our understanding of what it is to be an immigrant and how immigrants have managed to find success in the United States. For a qualitative approach to immigrants and immigration, Joel Millman's book *The Other Americans* (1997) provides a picture of immigrants from various countries and their paths to successful lives in the United States. Millman's theme is spelled out in the subtitle to his book: *How Immigrants Renew Our Country, Our Economy, and Our Values.* The stories of individual immigrants, their successes, and their struggles are transmuted into the statistics of professional advance, homeownership gains, and increasing political participation that are treated numerically in the present book. Both approaches bring understanding to immigrant outcomes and immigrant impacts.

NOTE

1. Personal communication from Karen Woodrow-Lafield.

Bibliography

Abelman, N. and Lee, J. 1995 *Blue Dreams: Korean Americans and the Los Angles Riots.* Cambridge, MA: Harvard University Press.

Alba, R. D. 1995. *Ethnic Identity: The Transformation of White America.* New Haven, CT: Yale University Press.

Alba, R. D. and Logan, J. 1992. Assimilation and stratification in the homeownership patterns of racial and ethnic groups. *International Migration Review* 26: 1314–1341.

Alba, R. D. and Logan, J. 1993. Minority proximity to whites in suburbs: An individual-level analysis of segregation. *American Journal of Sociology* 98: 1388–1427.

Alba, R. D., Logan, J. and Stults, B. J. 2000. The changing neighborhood contexts of the immigrant metropolis. *Social Forces* 79: 587–621.

Alba, R. D. and Nee, N. 1997. Rethinking assimilation theory for a new era of assimilation. *International Migration Review* 31: 826–874.

Allen, J. and Turner, E. 1996. Spatial patterns of immigrant assimilation. *The Professional Geographer* 48: 140–155.

Allen, J. and Turner, E. 1997. *The Ethnic Quilt.* Northridge: California State University, Northridge, Department of Geography, Center for Geographical Studies.

Allen, J. and Turner, E. 2001. Bridging 1990 and 2000 Census race data: Fractional assignment of multiracial populations. *Population Research and Policy Review* 20: 513–533.

Allen, J. and Turner, E. 2002. *Changing Faces, Changing Places: Mapping Southern Californians.* Northridge: California State University, Northridge, Department of Geography, Center for Geographical Studies.

American Housing Survey. 1999. Washington, DC: U.S. Department of Housing and Urban Development.

Archdeacon, T. J. 1983. *Becoming American: An Ethnic History.* New York: Free Press.

Archer, W., Ling, D., and McGill, G. 1996. The effect of income and collateral constraints on residential mortgage terminations. *Regional Science and Urban Economics* 26: 235–261.

Bachu, A. and O'Connell, M. 2001. *Fertility of American Women: June 2000.* Current Population Reports, P20-543RV. Washington, DC: U.S. Bureau of the Census.

Barclay, P. 1995. *Inquiry into Income and Wealth.* York, UK: Rowntree Foundation.

Barnett, D. 1999. *Show me the money.* Center for Immigration Studies, Backgrounder, April.

Barone, M. 2001. *The New Americans: How the Melting Pot Can Work.* Washington, DC: Regnery Press.

Bass, L. and Casper, L. 2002. Differences in registering and voting between native-born and naturalized Americans. *Population Research and Policy Review,* 20: 483–511.

Bayor, R. H. 1978. *Neighbors in conflict: The Irish, Germans, Jews and Italians of New York City 1929–1941.* Baltimore: Johns Hopkins University Press.

Bean, F. and Bell-Rose, S. (Eds.). 1999. *Immigration and Opportunity: Race, Ethnicity, and Employment in the United States.* New York: Russell Sage Foundation.

Bean, F., Chapa, J., Berg, R., and Sowards, K. 1994. Educational and sociodemographic incorporation amongst Hispanic immigrants to the United States. In B. Edmonston and J. Passel (Eds.), *Immigration and Ethnicity: The Integration of America's Newest Arrivals,* pp. 73–100. Washington DC: Urban Institute Press.

Beck, R. 1994. *Re-charting America's Future: Responses to Arguments against Stabilizing U.S. Population and Limiting Immigration.* New York: Social Contract Press.

Berends, M., Chun, J., Schuyler, G., Starkly, S. and Briggs, R. 2002. *Challenges of Conflicting School Reforms: Effects of New American Schools in a High-Poverty District.* Santa Monica, CA: RAND, MR-1483-EDU.

Binational Study on Migration. 1998. *Migration between Mexico and the United States.* Washington, DC: U.S. Commission on Immigration Reform.

Booth, A., Crouter, A., and Landale, N. (Eds.). 1997. Immigration and the Family: Research and Policy on US Immigrants. Mahwah, NJ: Erlbaum.

Borjas, G. 1990. *Friends or Strangers: The Impact of Immigrants on the U.S. Economy.* New York: Basic Books.

Borjas, G. 1998. *Heaven's Door: Immigration Policy and the American Economy.* Princeton, NJ: Princeton University Press.

Borjas, G., Freeman, R., and Katz, L. 1997. *How Much Do Immigration and Trade Affect Labor Market Outcomes?* Brookings Papers on Economic Activity. Washington, DC: Brookings Institution Press.

Bourassa, S. 1994. Immigration and housing tenure choice in Australia. *Journal of Housing Research* 5: 117–137.

Bourne, L. 1996. Social polarization and spatial segregation: Changing income inequalities in Canadian cities. In R. J. Davies (Ed.), *Contemporary City Structuring: International Geographical Insights*, pp. 136–157. Cape Town: Society of South African Geographers.

Bourne, L. 1998. *Migration, Immigration and Social Sustainability: The Toronto Experience in Comparative Perspective.* Paper presented at the UNESCO–MOST Conference, Cape Town, South Africa.

Bouvier, L. 1992. *Fifty Million Californians.* Washington, DC: Center for Immigration Studies.

Bouvier, L. 1996. *Embracing America: A Look at Which Immigrants Become Citizens.* Washington, DC: Center for Immigration Studies, Center Paper 11.

Bouvier, L. 1998. *The Impact of Immigration on United States' Population Size: 1950 to 2050. Washington, DC: Negative Population Growth (NPG Forum Series).*

Bouvier, L. and Jenks, R. 1998. *Doctors and Nurses: A Demographic Profile.* Washington, DC: Center for Immigration Studies, Center Paper 14.

Boyd, M. 1998. Triumphant transitions: Socioeconomic achievements of the second generation in Canada. *International Migration Review* 32: 853–876.

Brimelow, P. 1995. *Alien Nation: Common Sense about America's Immigration Disaster.* New York: Random House.

Burnett, D. 1999. *Show Me the Money: How Government Funding Has Corrupted Refugee Resettlement.* Washington, DC: Center for Immigration Studies, Backgrounder, April.

Burnley, I. 1999. Levels of immigrant residential concentration in Sydney and their relationship with disadvantage. *Urban Studies* 36: 1295–1315.

Burnley, I., Murphy, P., and Fagan, R. 1997. *Immigration and Australian Cities.* Sydney, New South Wales, Australia: Federation Press.

Butcher, K. and Card, D. 1991. Immigration and Wages: Evidence from the 1980s. *American Economic Review* 81: 292–296.

Cain, B. and Kiewiet, D. R. 1986. *Minorities in California.* Pasadena: California Institute of Technology.

Calder, L. 1999. *Financing the American Dream: A Cultural History of Consumer Credit.* Princeton, NJ: Princeton University Press.

Callis, R. 1997. *Moving to America—Moving to Home Ownership.* Washington, DC: U.S. Bureau of the Census (H121-92-2).

Camarota, S. 1997. Five million illegal immigrants: An analysis of new INS numbers. *Immigration Review* 28: 1–4.

Camarota, S. 1998. *The Wages of Immigration: The Effect on Low-Skilled Labor Markets.* Washington, DC: Center for Immigration Studies, Center Paper Number 12.

Camarota, S. 2001. *Immigration from Mexico: Assessing the Impact on the United States.* Washington, DC: Center for Immigration Studies, Center Paper Number 19.

Card, D. 1999. The causal effect of education on earnings. In O. Ashenfelter and D. Card (Eds.), *Handbook of Labor Economics* (Vol. 3). Amsterdam: North-Holland.

Carrington, W. and Detragiache, E. 1998. *How Big is the Brain Drain?* Washington, DC: International Monetary Fund, Working Paper 98/102.

Castles, S. and Miller, M. 1993. *The Age of Migration: International Population Movements in the Modern World.* New York: Guilford Press.

Chavez, L. 1996. Hispanics and the American Dream. *Imprimis* 25: 1–4.

Cheng, L. and Yang, P. 1996. Asians: The "model minority" deconstructed. In R. Waldinger and M. Bozorgmher (Eds.), *Ethnic Los Angeles,* pp. 305–344. New York: Russell Sage Foundation.

Chiswick, B. 1978. The effect of Americanization on the earnings of foreign-born men. *Journal of Political Economy* 86: 897–921.

Cho, W. 1999. Naturalization, socialization, participation: Immigrants and (non) voting. *Journal of Politics* 61: 1140–1155.

Clark, W. A. V. 1995. The expert witness in unitary hearings: The six green factors and spatial demographic change. *Urban Geography* 16: 664–679.

Clark, W. A. V. 1998. *The California Cauldron: Immigration and the Fortunes of Local Communities.* New York: Guilford Press.

Clark, W. A. V., Deurloo, M.C., and Dieleman, F. 2000. Housing consumption and residential crowding in U.S. housing markets. *Journal of Urban Affairs* 22: 49–63.

Clark, W. A. V. and McNicholas, M. 1995. Re-examining economic and social polarization in a multi-ethnic metropolitan area: The case of Los Angeles. *Area* 28: 56–63.

Clark, W. A. V. and Morrison, P. 1995. Demographic foundations of political empowerment in multi-minority cities. *Demography* 32: 183–201.

Clark, W. A. V. and Mueller, M. 1988. Hispanic relocation and spatial assimilation: A case study. *Social Science Quarterly* 69: 468–475.

Clarke, S. and Gaile, G. 1998. *The Work of Cities.* Minneapolis: University of Minnesota Press.

Cleeland, N. 1998. Irvine grows as Chinese gateway; Schools, high-tech jobs are magnets creating a demographic shift. *Los Angeles Times* December 7, p. 1.

Cornelius, W., Martin, P., and Hollifield, J. (Eds.). 1994. *Controlling Immigration: A Global Perspective.* Stanford, CA: Stanford University Press.

Currie, J. and Thomas, D. 1997. *Does Head Start Help Hispanic Children?* Santa Monica, CA: RAND.

de la Garza, R. 1992. *Latino Voices: Mexican, Puerto Rican and Cuban Perspectives on American Politics.* Boulder, CO: Westview Press.

de la Garza, R. and DeSipio, L. 1994. The links between individuals and electoral institutions. In R. de la Garza, M. Menchaca, and L. DeSipio (Eds.), *Barrio Ballots: Latino Politics in the 1990 Elections,* pp. 1–49. Boulder, CO: Westview Press.

de la Garza, R., Menchaca, M., and DeSipio, L. (Eds.). 1994. *Barrio Ballots: Latino Politics in the 1990 Elections.* Boulder, CO: Westview Press.

DeSipio, L. 1996a. *Counting on the Latino Vote: Latinos as a New Electorate.* Charlottesville: University of Virginia Press.

DeSipio, L. 1996b. Making citizens or good citizens: Naturalization as a predictor of organizational and electoral behavior among Latino immigrants. *Hispanic Journal of Behavioral Sciences* 18: 194–213.

DeWind, J. and Kasinitz, P. 1997. Everything old is new again?: Processes and theories of immigrant incorporation. *International Migration Review* 31: 1096–1111.

The Diversity Project. 1991. Final Report. Berkeley California: Institute for the Study of Social Change, University of California.

Duleep, H. and Dowhan, D. 2002. Insights from longitudinal data on the earnings growth of U.S. foreign-born men. *Demography* 39(3): 485–506.

Duncan, B. and Duncan, O.D. 1968. Minorities and the process of stratification. *American Sociological Review* 33: 356–364.

Duncan, G. and Brooks-Gunn, J. (Eds.). 1997. *The Consequences of Growing Up Poor.* New York: Russell Sage Foundation.

Duncan, O. D. and Lieberson, S. 1959. Ethnic segregation and assimilation. *American Journal of Sociology* 64: 367–374.

Edmonston, B. and Passel, J. (Eds.). 1994. *Immigration and Ethnicity: The Integration of America's Newest Arrivals.* Washington, DC: Urban Institute Press.

Edmonston, B. and Smith, J. 1997. *The New Americans.* Washington, DC: National Research Council.

Elliott, A. and Grotto, J. 2001. New Immigrant Majority: 51 percent in Miami, Dade were born in other nations. *Miami Herald* November 20, p. 1A.

Ellis, M. 2000. Mark one or more: Counting and projection by race in U.S. Census 2000 and beyond. *Social and Cultural Geography* 1: 183–195.

Ellis, M. and Wright, R. 1998. When immigrants are not migrants: Counting arrivals of the foreign born using the U.S. Census. *International Migration Review* 32: 127–144.

Espenshade, T. (Ed.). 1997. *Keys to Successful Immigration: Implications of the New Jersey Experience.* Washington, DC: Urban Institute Press.

Fainstein, S., Gordon, I., and Harloe, M. 1992. *Divided Cities: New York and London in a Divided World.* Oxford, UK: Blackwell.

Fannie Mae. 1995. *Fannie Mae National Housing Survey: Immigrants, Homeownership, and the American Dream.* Washington, DC: Author (Federal National Mortgage Association).

Farley, R. 1999. Racial issues: Recent trends in residential patterns and intermarriage. In N. Smelser and J. Alexander (Eds.), *Diversity and its Discontents,* pp. 85–128. Princeton, NJ: Princeton University Press.

Federal Reserve Bank of San Francisco. 2000. *Economic Letter,* September 15.

Fielding, A. 1995. Migration and social change: A longitudinal study of the social mobility of immigrants in England and Wales. *European Journal of Population* 11: 107–121.

Fix, M. E. and Passel, J. 1994. *Immigration and Immigrants: Setting the Record Straight.* Washington, DC: Urban Institute Press.

Foner, N. 2000. *From Ellis Island to JFK: New York's Two Great Waves of Immigration.* New Haven, CT: Yale University Press.

Fong, E. and Shibuya, K. 2000. Suburbanization and home ownership: The spatial assimilation process in U.S. metropolitan areas. *Sociological Perspectives* 43: 137–157.

Frey, W. 1996. Immigration, domestic migration, and demographic balkanization in America: New evidence for the 1990s. *Population and Development Review* 22: 741–763.

Friedberg, R. and Hunt, J. 1995. The impact of immigrants on host country wages, employment and growth. *Journal of Economic Perspectives* 9(2): 23–44.

Galbraith, J. K. 1976. *The Affluent Society.* Boston: Houghton Mifflin.

Gall, S. and Gall, T. (Eds.). 1993. *Statistical Record of Asian Americans.* Detroit, MI: Gall Research.

Galster, G., Metzger, K., and Waite, R. 1999. Neighborhood opportunity structures and immigrants socio-economic advancement. *Journal of Housing Research* 10: 95–127.

Gans, H. 1997. Toward a reconciliation of assimilation and pluralism: The interplay of acculturation and ethnic retention. *International Migration Review* 31: 875–892.

Ganzeboom, H., Treiman, D., and Ultee, W. 1991. Comparative intergenerational stratification research: Three generations and beyond. *American Review of Sociology* 17: 277–302.

Garces, E., Thomas, D., and Currie, J. 2002. Longer-Term Effects of Head Start. *American Economic Review* 92: 999–1012.

Garvey, D. L. 1997. Immigrants' earnings and labor-market assimilation: A case study of New Jersey. In T. Espenshade (Ed.), *Keys to Successful Immigration: Implications of the New Jersey Experience,* pp. 291–336. Washington, DC: Urban Institute Press.

Glazer, N. 1988. *The New Immigration: A Challenge to American Society.* San Diego, CA: San Diego State University Press.

Glazer, N. 1993. Is assimilation dead? *Annals, American Association of Political and Social Science* 510: 122–136.

Glick-Schiller, N. 1999. Transmigrants and nation-states: Something old and something new in the U.S. immigrant experience. In C. Hirschman, P. Kasinitz, and J. DeWind (Eds.), *The Handbook of International Migration: The American Experience,* pp. 94–119. New York: Russell Sage Foundation.

Glick-Schiller, N., Basch, L., and Blanc-Szanton, C. 1992. *Towards a Transnational Perspective on Migration.* New York: New York Academy of Sciences.

Gober, P. 1999. Settlement dynamics and internal migration of the US foreign born. In K. Pandit and S. D. Withers (Eds.), *Migration and Restructuring in the United States: A Geographic Perspective.* Lanham, MD: Rowman & Littlefield.

Goldsmith, M. and Blakely, E. 1992. *Separate Scieties: Poverty and Inequality in U.S. cities.* Philadelphia: Temple University Press.

Gordon, M. 1964. *Assimilation in American Life.* New York: Oxford University Press.

Gore, W., Hughes, M., and Galle, O. 1983. *Overcrowding in the Household.* New York: Academic Press.

Gottschalk, P. 1997. Inequality, income growth, and mobility: The basic facts. *Journal of Economic Perspectives* 11: 21–40.

Guinier, L. 1993. The representation of minority interests: The question of single-member districts. *Cardozo Law Review* 14: 1135–1174.

Gutmann, A. and Thompson, D. (Eds.). 1996. *Democracy and Disagreement.* Cambridge, MA: Harvard University Press.

Hamnett, C. 1994. Social polarization in global cities: Theory and evidence. *Urban Studies* 31: 401–424.

Haub, C. 1997. *Statistics on Mexican Fertility.* Washington, DC: Population Reference Bureau.

Haurin, D., Hendershott, P., and Wachter, S. 1996. Wealth accumulation and housing choices of young households: An exploratory investigation. *Journal of Housing Research* 7: 33–57.

Haurin, D., Hendershott, P., and Wachter, S. 1997. Borrowing constraints and the tenure choice of young households. *Journal of Housing Research,* 8: 137–154.

Haveman, R. and Wolfe, B. 1995. The determinants of children's attainments: A review of methods and findings. *Journal of Economic Literature* 33: 1829–1878.

Henderson, J. and Ionnides, Y. 1985. Tenure choice and the demand for housing. *Economica* 53: 231–246.

Heskin, A. 1983. *Tenants and the American Dream.* New York: Praeger.

Higham, J. 1975. *Send These to Me: Jews and Other Immigrants in Urban America.* New York: Atheneum.

Hirschman, C. 1983. The melting pot reconsidered. *Annual Review of Sociology* 9: 397–423.

Hirschman, C., Kasinitz, P., and DeWind, J. (Eds.) 1999a. *The Handbook of International Migration: The American Experience.* New York: Russell Sage Foundation.

Hirschman, C., Kasinitz, P., and DeWind, J. 1999b. Immigrant adaptation, assimilation and incorporation. In C. Hirschman, P. Kasinitz, and J. DeWind (Eds.), *The Handbook of International Migration: The American Experience,* pp. 127–136. New York: Russell Sage Foundation.

Hochschild, J. 1995. *Facing Up to the American Dream: Race, Class, and the Soul of the Nation.* Princeton, NJ: Princeton University Press.

Hollifield, J. 1996. The political economy of immigration: Markets versus rights in Europe and the United States. *Schriften des Zentralinstituts für Fränkische Landeskunde und Allegemeine Regionalforschung an der Universität Erlungen–Nürnberg, Wirkungen von Migration auf Aufnehmen Gesellschaften,* pp. 59–86.

Isbister, J. 1998. Is America too white? In E. Sandman (Ed.), *What, Then, Is the American, This New Man?,* pp. 27–32. Washington, DC: Center for Immigration Studies, Center Paper 13.

James, F. J. 1995. Urban economies: Trends; forces and implications for the President's national urban policy. *Cityscape: A Journal of Policy Development and Research* 1: 67–123.

Johnston, S. J., Katimin, M., and Milczarski, W. J. 1997. Homeownership aspirations and experiences: Immigrant Koreans and Dominicans in Northern Queens, New York City. *Cityscape: A Journal of Policy Development and Research* 3: 63–90.

Junn, J. 1999. Participation in liberal democracy: The political assimilation of immigrants and ethnic minorities in the United States. *American Behavioral Scientist* 42: 1417–1438.

Kasinitz, P. 1992. *Caribbean New York: Black Immigrants and the Politics of Race.* Ithaca, NY: Cornell University Press.

Kerr, T. 1996. *Chasing after the American Dream.* Commack, NY: Nova Science.

Kotkin, J. 1999. Immigration: The new ethnic entrepreneurs. *Los Angeles Times* September 12, p. 1.

Kotkin, J. 2002. Immigrants cushion the economic fall. *Wall Street Journal* January 17, p. 5.

Kraly, E., Powers, M., and Seltzer, W. 1998. U.S. immigration policy and immigrant integration: Occupational mobility among the population legalizing under IRCA. In L. Tomesi and M. Powers (Eds.), *Immigration Today: Pastoral and Research Challenges,* pp. 82–106. Staten Island, NY: Center for Migration Studies.

Krivo, L. J. 1995. Immigrant characteristics and Hispanic–Anglo housing inequality. *Demography* 32: 599–615.

Krivo, L. J. 1996. Household constraint and household complexity in metropolitan America: Black and Spanish-origin minorities. *Urban Affairs Quarterly* 21: 389–409.

LaFranchi, H. 1999. Why chic Mexican sneaks into U.S. *Christian Science Monitor* April 22, p. 1.

Lee, S. 1999. *Quality Based Investigation of Immigrants Housing Consumption: 1980–1990.* Unpublished paper presented at the annual meeting of the Western Regional Science Association, Ojai, California, February 21–24.

Leigh, N. G. 1994. *Stemming Middle Class Decline: The Challenges to Economic Development Planning.* New Brunswick, NJ: Rutgers University, Center for Urban Policy Research.

Levy, F. 1998. *The New Dollars and Dreams: American Incomes and Economic Change.* New York: Russell Sage Foundation.

Levy, F. and Michel, R. 1986. An economic bust for the baby boom. *Challenge* March–April, pp. 33–39.

Ley, D. and Murphy, P. 2001. Immigration in gateway cities: Sydney and Vancouver in comparative perspective. *Progress in Planning* 55: Pt. 3.

Ley, D. and Smith, H. 2000. Relations between deprivation and immigrant groups in large Canadian cities. *Urban Studies* 37: 37–62.

Ley, D. and Tutchener, J. 2001. Immigration, globalization and house prices in Canada's gateway cities. *Housing Studies* 16: 199–223.

Li, W. 1998. Los Angeles' Chinese *ethnoburb*: From ethnic service center to global economy outpost. *Urban Geography* 19(6), 502–517.

Lieberson, S. 1963. *Ethnic Patterns in American Cities.* New York: Free Press.

Lieberson, S. 1985. Unhyphenated whites in the United States. *Ethnic and Racial Studies* 8: 159–180.

Lieberson, S. and Waters, M. C. 1985. Ethnic mixtures in the United States. *Sociology and Social Research* 70: 43–52.

Lieberson, S. and Waters, M. C. 1988. *From Many Strands: Ethnic and Racial Groups in Contemporary America.* New York: Russell Sage Foundation.

Light, I. and Bhachu, P. (Eds.). 1993. *Immigration and Entrepreneurship: Culture, Capital, and Ethnic Networks.* New Brunswick, NJ: Transaction Books.

Light, I. and Roach, E. 1996. Self-employment: Mobility ladder or economic lifeboat? In R. Waldinger and M. Bozorghmer (Eds.), *Ethnic Los Angeles,* pp. 193–213. New York: Russell Sage Foundation.

Lipset, S. M. and Raab, E. 1970. *The Politics of Unreason: Right Wing Extremism in America, 1790–1970.* New York: Harper Torchbook.

Loewen, J. 1988. *The Mississippi Chinese.* Prospect Heights, IL: Waveland Press.

Logan, J. R., Alba, R., and McNulty, T. 1994. Ethnic economies in metropolitan regions: Miami and beyond. *Social Sciences* 72: 691–724.

Lollock, L. 2001. *The Foreign-Born Population in the United States: March 2000.* Washington, DC: U.S. Bureau of the Census, Current Population Reports, P20-534.

Lopez, D. 1996. Language, diversity and assimilation. In R. Waldinger and M. Bozorghmer (Eds.), *Ethnic Los Angeles,* pp. 139–163. New York: Russell Sage Foundation.

Lopez, E., Ramirez, E., and Rochin, R. 1999. *Latinos and Economic Development in California.* Sacramento: California State Library, California Research Bureau, CRB99-008.

Martin, J. 1995. A surge in naturalization. *Immigration Review* 22: 9.

Martin, P. and E. Midgley, E. 1994. Immigration to the United States: Journey to an uncertain destination, *Population Bulletin* 49: 1–47.

Masnick, G. S., Pitkin, J. R., and Brennan, J. 1990. Cohort housing trends in a local housing market: The case of Southern California. In D. Myers (Ed.), *Housing Demography,* Madison: University of Wisconsin Press.

Massachusetts Insight. 1999. *Made in Massachusetts: Competitive Manufacturing in a High-Skill Location.* Boston: Massachusetts Corporation.

Massey, D. S. and Espinosa, K. E. 1997. What's driving Mexico–U.S. migration?: A theoretical, empirical, and policy analysis. *American Journal of Sociology* 102: 939–99.

Massey, D. S., Goldring, L., and Durand, J. 1994. Continuities in transnational migration: An analysis of nineteen Mexican communities. *American Journal of Sociology,* 99(6): 1492–1533.

McArdle, N. 1997. Homeownership attainment of New Jersey immigrants. In T. Espenshade (Ed.), *Keys to Successful Immigration: Implications of the New Jersey Experience,* pp. 337–369. Washington, DC: Urban Institute Press.

McCarthy, G., Van Zandt, S., and Rohe, W. 2001. *The Economic Benefits and Costs of Homeownership: A Critical Assessment of the Research.* Arlington, VA: Research Institute for Housing America.

McCarthy, K. and Vernez, G. 1997. *Immigration in a Changing Economy: The California Experience.* Santa Monica, CA: RAND.

McHugh, K. 1989. Hispanic migration and population redistribution in the United States. *Professional Geographer* 41: 429–439.

McHugh, K., Miyares, I., and Skop, E. 1997. The magnetism of Miami: Segmented paths in Cuban migration. *Geographical Review* 87: 504–519.

McWillaims, C. 1973. *Southern California: An Island on the Land.* Santa Barbara, CA: Peregrine Smith.

Miele, S. 1920. America as a place to make money. *World's Work* 41: 204.

Miller, D. C. 1991. *Handbook of Research Design and Social Measurement.* Newbury Park, CA: Sage.

Millman, J. 1997. *The Other Americans: How Immigrants Renew Our Country, Our Economy, and Our Values.* New York: Viking.

Model, S. 1985. A comparative perspective on the ethnic enclave: Blacks, Italians, and Jews in New York City. *International Migration Review* 19: 64–81.

Moore, E. and Rosenberg, M. 1995. Modeling migration flows of immigrant groups in Canada. *Environment and Planning* A27: 699–714.

Morrison, P. 1974. Urban growth and decline: San Jose and St. Louis in the 1960's. *Science* 185: 757–762.

Morrison, P. 1998. Demographic influences on Latinos' political empowerment: Comparative local illustrations. *Population Research and Policy Review* 17: 223–246.

Morrison, P. 2000. Forecasting enrollments for immigrant entry-port school districts. *Demography* 37: 499–510.

Morrison, P. and Wheeler, J. 1978. The image of "elsewhere" in the American tradition of migration. In W. H. McNeill and R. Adams (Eds.), *Human Migration: Patterns and Policies,* pp. 75–84. Bloomington: Indiana University Press.

Muller, T. 1993. *Immigrants and the American City.* New York: New York University Press.

Myers, D. 1985. Wives earnings and rising costs of home ownership. *Social Science Quarterly* 66: 319–329.

Myers, D. 1999. Cohort longitudinal estimation of housing careers. *Housing Studies* 14: 473–490.

Myers, D., Baer, W., and Choi, S. 1996. The changing problem of overcrowded housing. *Journal of the American Planning Association* 62: 66–96.

Myers, D. and Crawford, C. 1998. Temporal differentiation in the occupational mobility of immigrant and native-born Latina workers. *American Sociological Review* 63: 68–93.

Myers, D. and Lee, S. W. 1996. Immigration cohorts and residential overcrowding in Southern California. *Demography* 33: 51–65.

Myers, D. and Lee, S. W. 1998. Immigrant trajectories into homeownership: A temporal analysis of residential assimilation. *International Migration Review* 32: 595–625.

Myers, D., Megbolugbe, I., and Lee, S. W. 1998. Cohort estimation of homeownership attainment among native-born and immigrant populations. *Journal of Housing Research* 9: 237–269.

Myers, D. and Park, J. 1999. The role of occupational achievement in homeownership attainment by immigrants and native borns in five metropolitan areas. *Journal of Housing Research* 10: 61–93.

NALEO Educational Fund. 1989. *The National Latino Immigrant Survey.* Washington, DC: National Association of Latino Elected and Appointed Officials.

NALEO Educational Fund. 1992. *National Register of Hispanic Elected Officials.* Los Angeles: National Association of Latino Elected and Appointed Officials.

National Association of Realtors. 1992. *Survey of Homeowners and Renters.* Washington, DC: Author.

Neidert, L. J., and Farley, R. 1985. Assimilation in the United States: An analysis of ethnic and generational differences in status and achievement. *American Sociological Review* 50: 840–850.

Nelli, H. S. 1983. *From Immigrants to Ethnics: The Italian Americans.* Oxford, UK: Oxford University Press.

Newman, K. 1993. *Declining Fortunes: The Withering of the American Dream.* New York: Basic Books.

O'Connor, A. M. 1999. Roar of Soccer at Coliseum. *Los Angeles Times,* April 29, p. 1.

Olzak, S. 1992. *The Dynamics of Ethnic Competition and Conflict.* Stanford, CA: Stanford University Press.

Ong, P., Bonacich, E., and Cheng, L. (Eds.), 1994. *The New Asian Immigration in Los Angeles and Global Restructuring.* Philadelphia: Temple University Press.

Ong, P. and Hee, S. 1994. Economic diversity. In P. Ong, E. Bonacich, and L. Cheng (Eds.), *The Asian Immigration in Los Angeles and Global Restructuring,* pp. 31–53. Philadelphia: Temple University Press.

Oropesa, R. S. and Landale, M. S. 1997. Immigrant legacies: Ethnicity, generation and children's familial and economic lives. *Social Science Quarterly* 78: 399–416.

Ortiz, V. 1996. The Mexican-origin population: Permanent working class or emerging middle class? In R. Waldinger and M. Bozorghmer (Eds.), *Ethnic Los Angeles,* pp. 247–277. New York: Russell Sage Foundation.

Pader, E. 1994. Spatial relations and housing policy: Regulations that discriminate against Mexican-origin households. *Journal of Planning Education and Research* 13: 119–135.

Painter, G. Gabriel, S., and Myers, D. 2000. *The Decision to Own: The Impact of Race, Ethnicity and Immigrant Status.* Arlington, VA: Research Institute for Housing America.

Parker, R. 1972. *The Myth of the Middle Class.* New York: Liveright.

Partridge, M., Levernier, W., and Rickman, D. 1996. Trends in US income inequality: Evidence from a panel of states. *Quarterly Review of Economics and Finance* 36: 17–37.

Passel, J. 2001. *Demographic Analysis: An Evaluation,* pp. 86–113. Washington, DC: Report to the U.S. Census Monitoring Board.

Perlman, J. 1997. *Reflecting the changing face of America: Multiracials, racial classification, and American intermarriage.* New York: The Jerome Levy Economics Institute of Bard College, Public Policy Brief 35.

Phillips, A. 1994. *Democracy and Difference.* University Park: Pennsylvania State University Press.

Pickus, N. M. 1998. To make natural: Creating citizens for the twenty-first century. In N. M. Pickus (Ed.), *Immigration and Citizenship in the 21st Century.* pp. 107–139. New York: Rowman & Littlefield.

Pigeon, M. and Wray, L. R. 1998. *Did the Clinton rising tide raise all boats?* Public policy brief, no. 45, Annandale-on-Hudson, NY: Bard College, Jerome Levy Economics Institute.

Plotke, D. 1997. Immigration and political incorporation in the contemporary United States. In C. Hirschman, P. Kasinitz, and J. DeWind (Eds.), *The Handbook of International Migration: The American Experience,* pp. 294–318. New York: Russell Sage Foundation.

Portes, A. and Rumbaut, R. 1996. *Immigrant America: A Portrait.* Berkeley: University of California Press.

Portes, A. and Stepick, A. 1993. *City on the Edge: The Transformation of Miami.* Berkeley: University of California Press.

Portes, A. and Zhou, M. 1992. Gaining the upper hand: Economic mobility among immigrant and domestic minorities. *Ethnic and Racial Studies* 15: 491–523.

Portes, A. and Zhou, M. 1993. The new second generation: Segmented assimilation and its variants. *Annals of the American Academy of Political and Social Science* 530: 74–96.

Portes, A. and Zhou, M. 1994. Should immigrants assimilate? *The Public Interest* 116: 18–33.

Powers, M. and Seltzer, W. 1998. Occupational status and mobility among undocumented immigrants by gender. *International Migration Review* 32(1): 21–55.

Powers, M., Seltzer, W., and Shi, J. 1998. Gender differences in the occupational status of undocumented immigrants in the United States: Experience before and after legalization. *International Migration Review* 32: 1015–1046.

Price, M. 2002. *The Creation of Latino Cultural Space through Soccer: The Futbal Leagues of Washington, D.C.* Unpublished paper presented at the annual meeting of the Association of American Geographers, Los Angeles, CA.

Ramakrishnan, S. K. 2001. *The Political Incorporation of Immigrants and Their Descendants.* Unpublished paper presented at the meetings of the Population Association of America, Washington, DC.

Ramakrishnan, S. K. and Espenshade, T. 2000. *Immigrant Incorporation and Political Participation: Generational Status and Voting Behavior in U.S. Elections.* Unpublished paper presented at the annual meeting of the Population Association of America, Los Angeles, CA.

Ratner, M. S. 1996. Many routes to homeownership: A four site ethnographic study of minority and immigrant experiences. *Housing Policy Debate* 7: 103–145.

Ravitch, D. 1990. Multiculturalism: *E Pluribus Plures. American Scholar* 59: 337–354.

Reddy, N. (Ed.). 1993. *Statistical Record of Hispanic Americans.* Detroit, MI: Gale Research Inc.

Reed, D. 1999. *California's Rising Income Inequality: Causes and Concerns.* San Francisco: Public Policy Institute of California.

Reed, D. 2001. Immigration and males' earnings inequality in the regions of the United States. *Demography* 38: 363–373.

Reich, R. 1991. *The Work of Nations: Preparing Ourselves for 21st Century Capitalism.* New York: Knopf.

Reisman, D. 1980. *Galbraith and Market Capitalism.* New York: New York University Press.

Reitz, J. 1997. Terms of entry: Social institution and immigrant earnings in America, Canadian and Australian cities. In M. Cross and R. Moore (Eds.), *Globalization and the New City: Migrants, Minorities and Urban Transformations in Comparative Perspective,* pp. 50–81. Houndsmill, Basingstoke, Hampshire, UK: Palgrave.

Reitz, J. 1998. *The Warmth of the Welcome: The Social Causes of Economic Success for Immigrants in Different Nations and Cities.* Boulder, CO: Westview Press.

Renshon, S. A. 2001.*Citizenship and American National Identity.* Washington, DC: Center for Immigration Studies.

Rich, M. 1993. *Federal Policy Making and the Poor: National Goals, Local Choices and Distributional Outcomes.* Princeton, NJ: Princeton University Press.

Rodriguez, G. 1996. *The Emerging Latino Middle Class.* Los Angeles: Pepperdine University.

Rodriguez, G. 1999. *From Newcomers to New Americans.* Washington, DC: National Immigration Forum.

Rolph. E. 1992. *Immigration Policies: Legacy from the 1980's and Issues for the 1990's.* Santa Monica, CA: RAND.

Rosenfeld, M. [J.] 1998. Mexican immigrants and Mexican American political assimilation. In Binational Study on Migration, *Migration between Mexico and the United States*, pp. 1117–1131. Washington, DC: U.S. Commission on Immigration Reform.

Rosenfeld, M. J. and Tienda, M. 1999. Mexican immigration, occupational niches, and labor-market competition: Evidence from Los Angeles, Chicago, and Atlanta, 1970–1990. In F. D. Bean and S. Bell-Rose (Eds.), *Immigration and Opportunity*, pp.64–105. New York: Russell Sage Foundation.

Rumbaut, R. 1997. *Passages to Adulthood: The Adaptation of Children of Immigrants in Southern California.* New York: Report to the Russell Sage Foundation.

Sassen, S. 1994. *Cities in a World Economy.* London: Pine Forge Press.

Saxenian, A. 1999. *Silicon Valley's New Immigrant Entrepreneurs.* San Francisco: Public Policy Institute of California.

Schoeni, R. 1998a. Labor market outcomes of immigrant women in the United States: 1970 to 1990. *International Migration Review* 32: 57–73.

Schoeni, R. 1998b. Labor market assimilation of immigrant women. *Industrial and Labor Relations Review* 51: 483–504.

Simon, J. 1989. *The Economic Consequences of Immigration.* Cambridge, MA: Blackwell.

Simon, J. and Akburi, A. 1995. *Educational Trends of Immigrants in the U.S.* Unpublished paper presented at the annual meeting of the Population Association of America, Washington, DC.

Skerry, P. 1993. *Mexican Americans: The Ambivalent Minority.* New York: Free Press.

Skerry, P. 1998. Do We Really Want Immigrants to Assimilate? In E. Sandman (Ed.), *What, Then, Is the American, This New Man?*, pp. 39–48. Washington, DC: Center for Immigration Studies, Center Paper 13.

Smith, J. and Edmonston, B. 1997. *The New Americans: Economic, Demographic and Fiscal Impacts of Immigration.* Washington, DC: National Academy Press.

Sosa, L. 1998. *The Americano Dream: How Latinos Can Achieve Success in Business and in Life.* New York: Dutton.

Stasz, C., Chiesa, J., and Shwabe, W. 1998. *Education and the New Economy: A Policy Planning Exercise.* Berkeley: University of California, Berkeley, National Center for Research in Vocational Education.

Suro, R. 1994. *Remembering the American Dream: Hispanic Immigration and National Policy.* New York: Twentieth Century Fund Press.

Suro, R. and Singer, A. 2002. *Latino Growth in Metropolitan America: Changing Patterns, New Locations.* Washington, DC: Brookings Institution, Center on Urban and Metropolitan Policy.

Taylor, E. and Martin, P. 1997. *Poverty amid Prosperity: Immigration and the Changing Face of Rural California.* Washington, DC: Urban Institute.

Thernstrom, A. 1987. *Whose Votes Count: Affirmative Action and Minority Voting Rights.* Cambridge, MA: Harvard University Press.

Tienda, M. 1999. Immigration, diversity and equality of opportunity. In N. Smelser and J. Alexander (Eds.), *Diversity and Its Discontents*, pp. 129–146. Princeton, NJ: Princeton University Press.

Tobar, H. 1999. Living *la Vida Buena* in Georgia. *Los Angeles Times*, December 29, p. 1.

Topolnicki, D. 1991. Why we still live best.*Money* October, pp. 86–92.

Torres-Gil, F. 1998. *The Politics of Longevity and Diversity: Implications and Lessons from the Aging of the United States Population.* Los Angeles: University of California, Los Angeles, School of Public Policy.

Treiman, D. and Ganzeboom, H. 1990. Comparative status attainment research. *Research Social Spatial Mobility* 9: 105–127.

Uhlaner, C., Cain, B., and Kiewiet, R. 1989. Political participation of ethnic minorities in the 1980s. *Political Behavior* 11: 195–231.

Urban Institute. 2000. *Checkpoints, data releases on economic and social issues.* Septermber 2, p. 7: Washington, DC.

U.S. Bureau of the Census. 1972. *U.S. Census of Population and Housing, 1970.* Washington, DC: Author.

U.S. Bureau of the Census. 1983. *U.S. Census of Population and Housing, 1980, Public Use Microdata Samples (California).* Washington, DC: Author.

U.S. Bureau of the Census. 1991. *U.S. Census of Population and Housing, 1990, Summary Tape File 1 (California).* Washington, DC: Author.

U.S. Bureau of the Census. 1992. *U.S. Census of Population and Housing, 1990, Public Use Microdata Samples (California).* Washington, DC: Author.

U.S. Bureau of the Census. 1993. *U.S. Census of Population and Housing, 1990, Summary Tape File 3 (California).* Washington, DC: Author.

U.S. Bureau of the Census. 1998. *Money Income in the United States.* Current Population Reports, P60-200. Washington, DC: Author.

U.S. Bureau of the Census. 1999. *American Housing Survey for the Los Angeles–Long Beach Metropolitan Area.* Washington, DC: U.S. Government Printing Office.

U.S. Bureau of the Census. 2000. *Current Population Survey.* Ann Arbor, Michigan: Interuniversity Consortium for Political and Social Research.

U.S. Bureau of the Census. 2001. *Current Population Survey.* Ann Arbor, Michigan: Interuniversity Consortium for Political and Social Research.

U.S. Commission on Immigration Reform. 1997. *Becoming an American: Immigration and Immigrant Policy.* Washington, DC: U.S. Government Printing Office.

U.S. Department of Labor. 1989. *The Effects of Immigration on the U.S. Economy and Labor Market.* Washington, DC: U.S. Government Printing Office.

U.S. Immigration and Naturalization Service. 1996. *Statistical Yearbook of the Immigration and Naturalization Service.* Washington, DC: U.S. Government Printing Office.

U.S. Immigration and Naturalization Service. 1999. *Statistical Yearbook of the Immigration and Naturalization Service.* Washington, DC: U.S. Government Printing Office.

Vernez, G., Krop, R., and Rydell, P. 1999. *Closing the Education Gap: Benefits and Costs.* Santa Monica, CA: RAND, MR-1036-EDU.

Waldinger, R. D. and Bozorgmehr, M. (Eds.). 1996. *Ethnic Los Angeles.* New York: Russell Sage Foundation.

Waldinger, R. D. (Ed.) 2001. *Strangers at the Gates.* Berkeley: University of California Press.

Warren, R. 2000. *Annual Estimates of the Unauthorized Immigrant Population Residing in the United States and Components of Change: 1987 to 1997.* Unpublished document. Washington, DC: U.S. Immigration and Naturalization Service.

Weintraub, S., Alba, F., Fernández de Castro, R., and García y Griego, M. 1998. Responses to migration issues. In Binational Study on Migration, *Migration between Mexico and the United States,* pp. 437–509. Washington, DC: U.S. Commission on Immigration Reform.

Wolfinger, R. and Rosenstone, S. 1980. *Who Votes?* New Haven, CT: Yale University Press.

Woodrow-Lafield, K., Xu, X., Kersen, K., and Poch, B. 2001. Naturalization experiences among U.S. immigrants. Proceedings of the Social/Government Statistics Section, American Statistical Association. In *Joint Statistical Meetings,* May, pp. 106–111.

Wright, R., Bailey, A., Miyares, I., and Mountz, A. 2000. Legal status, gender and employment among Salvadorans in the U.S. *International Journal of Population Geography* 6: 273–286.

Wright, R. and Ellis, M. 1996. Immigrants and the changing racial/ethnic division of labor in New York City. *Urban Geography* 17: 317–353.

Wright, R. and Ellis, M. 1997. Nativity, ethnicity, and the evaluation of the intraurban division of labor in metropolitan Los Angeles 1970–1990. *Urban Geography* 18: 243–263.

Zhou, M. 1997. Segmented assimilation: Issues, controversies, and recent research on the new second generation. *International Migration Review* 31: 975–1008.

Index

Acculturation, 13, 14, 157
Achievement, 208
Africa, 124
professionals, 111
African Americans, 16, 20, 60, 166, 170, 186–187, 191, 220
Age, 40, 78, 152–153, 159, 175, 178, 180, 201
and homeownership, 18, 141–144
naturalization, 176
pyramids, 41–42
Agriculture, 101, 214
farm workers, 104, 176–177, 179
American Dream, 1–28, 62, 155, 199, 208
and assimilation, 12
definition, 6
integration, 15
material, 4–5, 9
mythology, 4, 199
rootedness, 19
success, 4, 17, 25, 45
American movies, 10
Americanization, 163
Armenian, 213
Asia, 15, 31, 34, 41
Asians, 18, 149, 169–171, 182–183, 203–205, 214
elected officials, 173
entrenpreneurs, 12, 109–110
homeownership, 135, 137–142, 144–145, 150–154, 156
language use, 45
middle class, 13, 64–68, 72–77, 82–90
model minority, 71
professional occupation, 117
Assimilation, 3, 12–14, 17, 22, 51–52, 62–63
blending, 12, 165–167, 169, 191, 222
citizenship, 157, 160, 163–167, 169–170
cultural, 163
economic, 76, 163
homeownership, 127, 129–130, 145–146, 151
incorporation, 13, 50–51
integration, 18
linear, 14, 165
naturalization, 175
political participation, 191
segmented, 12–13, 18
voting, 184, 188
Asylum seekers, 222
Austin TX, 111

B

Baby boomers, 198, 221
Bilingual, 24, 44–45
Bill of rights, 2
Bio technology, 111
Births (*see also* Fertility), 30, 220
Blue collar workers, 104–105, 172, 177, 179
Borders, 56, 146, 222

C

California, 10, 21, 24, 26, 31, 39, 167, 187, 197, 201, 223
age, 42
cities, 148
ethnic diversity
homeownership, 126, 132–135, 137–143, 148, 158–159

California *(cont.)*
  immigrant concentration, 36–37
  income inequality, 202–205
  labor force, 100, 102–103
  middle class, 74–75, 80–91
  naturalization, 174
  professional occupations, 113–119, 121, 123
Cambodians, 45, 96
Canada (Canadians), 49, 100, 108, 112, 117, 133, 151, 171
  cities, 146–147
  voting, 182
Capital, 56, 129
Caribbean, 43, 48
Census, 11, 30–31, 39, 157, 172, 175
  data, 225, 229
Central America(ns), 15, 19–20, 32, 34, 40, 50, 57, 151, 219
  homeownership, 151
  professional occupations, 108, 112, 118
  voting, 179–180
Central cities, 155–157, 171
Chain migration, 45
Childbearing, 40
Children, 76, 167, 208–209, 210, 212
  school age, 42, 44
China (Chinese), 10, 15, 19, 36, 42, 49, 100, 169, 180
  earnings, 49
  homeownership, 151
  professional occupations, 108–110, 112–118
  voting, 182
Cities, 44, 57, 146–147, 149
Citizenship, 19, 78–79, 145, 163–165, 211, 223
  and assimilation, 169
  homeownership, 158–160
  naturalization, 170, 172–173
Civic life, 164
Civil rights, 16, 164, 169–170
Cohorts, 17, 97, 130, 223
  middle class, 70–76
  homeownership, 131–134, 136–15
College education, 46, 48, 125, 159, 176, 217
Communities, 22, 24–25, 45, 50, 57, 80, 160, 169, 198, 191
  housing, 129, 130, 146–148
  participation, 172
Communities of interest, 189, 191
Commuting, 157–158
Conflict, 168
Congestion, 145
Congressional districts, 189–190
Connecticut, 32
Crowding, 145, 148
Cuba (Cubans), 36, 100, 108, 112, 214
  and homeownership, 133–134
  naturalization, 178–179
  professional occupations, 117
  voting, 182
Culture(s), 165
Cultural heritage, 11, 167
Cultural landscape, 24
Current Population Survey, 47

D

Dallas, 45, 210
Data
  American Housing Survey, 149
  Census, 225, 229
  Current Population Survey, 225–227, 230
  Immigration and Naturalization Service, 225, 228
  Mexican Migration Project, 226
  Public Use Microdata Sample (PUMS), 226
  qualitative, 230
Democracy, 5, 19, 173
Democrats, 189
Democratic society, 5, 164, 171
Dependency (see Welfare)
Diversity, 34, 149, 167
Discrimination, 4, 12, 128, 145, 166, 170, 187, 189
Doctors (see occupations)
Dominican Republic (Dominicans), 49, 133, 140, 182, 214
Dot coms, 2
Dual citizenship, 167, 169, 211
Dual economy, 195

E

Earnings (*see also* Income), 61, 131
  education, 58
  Central Americans, 58
  Mexicans, 58
  foreign born, 49
  native born, 61
  recent immigrants, 49
Eastern Europe, 22, 42, 82, 117

Economy/Economic contexts, 89–91, 98, 199
Economic growth, 128, 209, 216
Economic Status, 17
Education, 1, 46–48, 52, 62–63, 77–78, 111, 114, 210
attainment, 48, 51, 95, 130, 177
children, 177, 215–217
college (degrees), 1, 26, 47–48, 60, 107, 111–112, 159, 197
community colleges, 112
funding, 217
gap, 48, 215
naturalization, 174, 176–177
voting, 179–181
Elections
districts, 166, 172, 186–187
politics, 172–173, 189
Electoral outcomes, 186–188
El Salvador, 26, 96
Employment, 15, 196, 212, 215
opportunities, 48
Enclaves, 57, 145
Engineers, 105
English language, 78–79, 129, 131, 165, 174–175, 183, 211
Entrepreneurs, 19, 97, 109–110, 213
Entry ports, 21, 146
Ethnic, 169
groups, 166, 172
districts, 186–187
enclaves, 57, 145
status, 229, 229
Ethnicity, 201, 223
Europe (European), 34–35, 133, 151, 174

F
Family reunification, 20, 54
Family contacts, 26
Farmworkers, 104, 176–177, 181
Fertility, 219–220
age-sex pyramids, 41–41
births, 30, 200
rates, 147
Filipinos, 107, 122, 214
First Generation, 3, 130, 167–168, 182
Florida, 21, 31, 37, 39, 201, 203–205, 223
homeownership, 132–134, 137–143, 158–159
labor force, 100, 102–103
middle class, 64–69, 74–75, 80–91
professional occupations, 113–119, 121, 123
Foreign born, 29–39, 42, 57
children, 43
labor force, 98–103
France (French), 5–6

G
Garment industry, 110
Gateway cities, 23, 147–155, 208
Gender, 115
Generations, 76–77
Germany (Germans), 5–6, 12, 14, 16
Glendale, 213
Global cities, 146
Globalization, 55–56, 195–196
Great Britain (England), 5–6

H
H-1B visa, 55
Hart-Cellar Act, 15, 54
Head Start, 215
Health care, 24, 209, 220
High school, 215
drop-outs, 218
High tech, 19, 109–110, 112, 195
Hispanic, 11, 129, 149, 166, 169–170, 200
elected officials, 173
homeownership, 130, 135, 137–142, 144, 150–154
inequality, 203–205
middle class, 64–68, 72–77, 82–90
naturalization, 170–171, 178–179
professional occupation, 117
voting, 182–183, 189
Hmong, 45
Homebuying, 126–127
Homeownership, 1, 8, 19, 62–63, 78, 83, 128, 179, 198
age, 141–144, 158
affordability, 126, 147, 149
Asians, 135, 138–141, 144
citizenship, 179
costs, 128, 145, 149
determinants, 158–160
equity, 129
financial advantages, 8
Hispanics, 130, 135, 137–142, 144, 150–154, 156
investment, 129
Mexicans, 130
Non-Hispanic white, 135, 138–141, 144
ownership transitions, 145

Homeownership *(cont.)*
prices, 136, 145–147, 149, 155
rates, 137–145, 150–154, 199
suburban, 145, 155–158, 160
time of entry, 135, 154, 158–159
trajectories, 131–134, 136, 158
Hong Kong, 19
Household size, 43
Housing
affordability, 149, 155, 158, 160
crowding, 148
equity, 129
markets, 127–128, 145, 148
quality, 128, 145
values, 149
prices, 155
Human capital, 19, 21, 25–26, 58, 62, 109, 111, 124, 211, 216
and education, 45–46
homeownership, 157–158, 160
inequality, 195

I

Illegal immigration (see Undocumented migrants)
Immigrant(s)
advocacy groups, 16
concentrations, 38–39, 45, 80–81, 216
definitions, 225
entrepreneurs, 38, 97, 109–110
flows, 53
homeownership, 126–160
legal, 15, 20
low skill, 111, 176, 218, 220
illegal, 20, 165
networks, 16
pioneers, 10
progress, 60
rights, 16
success, 27, 61
undocumented, 15, 53, 124, 168, 172, 175
visas, 55
Immigration
Acts, 54–55
bifurcated, 19–21, 47, 55, 94
channelized, 45
composition, 57
controls, 53
demand pull, 15
economic opportunities, 210
family reunification, 54
flows, 1, 2, 14–15, 21, 34, 52, 55, 165
historical, 30
labor flows, 56, 99
networks, 16
policy, 53–55, 222
self selection, 10
waiting list, 54
IRCA, 53–54, 136
Income (*see also* Earnings), 6, 94, 130, 158, 177–178, 199–201
deciles, 148
distribution, 205
inequality, 195–198, 200–206
low, 144, 148, 178
middle, 7, 127, 144, 178, 197, 211
quintiles, 18, 198
ranges, 7, 8, 18
squeeze, 198
two earners, 9, 198
upper, 144, 177–178, 201
India (Indians), 19, 36, 49–50, 100, 151, 182
middle class, 79
professional occupations, 108–109, 111–112, 118
Incorporation, 14, 50, 163–165, 170
Integration, 22, 129, 156, 160, 163, 179, 199
Inter-marriage, 22, 164, 169, 223
Industry, 94–101
Jewelry, 110
Iran, 211, 213
Irish, 169
Italy (Italitans), 3, 11–12, 14, 169, 172

J

Japan (Japanese), 96, 141, 182
Jews, 11, 14, 35, 169
Jobs, 157
competition, 217–218
construction, 105
finance/business, 105
low-pay, 20
low skill, 48, 195
manufacturing, 105
public administration, 105

K

Korea (Koreans), 19, 22, 24, 36, 49, 180
citizenship, 79
earnings, 49
entrepreneurs, 97
housing, 151
middle class, 78–79

naturalization, 179
professional occupation, 108
voting, 182

L

Laborforce
growth, 100–103
composition, 108
women, 119–122
Language, 22, 31, 33, 78, 145, 166, 168, 185
Laos (Laotian), 19, 45
Las Vegas, 21–22
Latino(s) (*see also* Hispanic), 11, 12, 18, 20, 62, 166, 171
age pyramid, 41–42
education, 215
politcal, 186–189
Leisure, 9
Linguistic isolation, 45
Loans, 26
Los Angeles, 13, 15, 22–25, 38, 89, 149–155, 157, 211, 214
Los Angeles Times, 10, 22, 26, 111, 157, 189
Low tech, 2
Lowell (Mass.), 45
Loyalty, 167–168

M

Market economy, 5
Material wellbeing, 2
Manual laborers, 171
Manufacturing jobs, 195
Media, 3, 10, 55–56, 64, 165, 169, 221
Melting pot, 12, 35, 165, 191
Metropolitan, 157
Mexicans, 40, 56, 213
age distribution, 42
communities, 16
earnings, 50
education, 218
home ownership, 136
laborforce, 99, 108
middle class, 79, 85
naturalization, 179
professional occupation, 96, 112, 118
voting, 179–180
Mexico, 10, 15, 16, 19, 32, 34, 36, 57, 133, 182
dual citizenship, 168
population, 219
wages, 209
Miami, 39, 57, 214
Middle class, 11, 13, 22, 60, 109, 122, 126, 194, 198, 222
and American dream, 24–27
cohort changes, 70–76
definition, 6, 9, 62–63
foreign born, 63–69, 80–89
growth, 65, 86
lower/upper, 7–8, 202
native born, 63–69, 80–89
norms, 13,
numbers, 64–65, 81–83
probability of access, 78–79
progress, 17–19, 208–214
proportions, 66–69, 83–89, 91
social integration, 61
status, 6, 8
time of entry, 89–91
vanishing, 61
Middle Eastern, 18, 35, 169
home ownership, 151
professional occupation, 108–110, 118
voting, 179–180
Middle income, 84, 197–198
Midwest, 38–40, 44
Migrants (see Immigrants)
Migration (see Immigration)
Mississippi, 15
Mortgage, 127, 129, 210
Multiculturalism, 164, 167
Multi-ethnic (multiracial), 189

N

Nation (state), 164
National Academy Sciences (The New Americans), 51, 218
National Assn. Realtors, 129
National identity, 164
National quotas, 54
Native born laborforce, 98–103
Naturalization, 131, 163–165, 170–173, 185
time of arrival, 175–176
homeownership, 179
Neighborhoods, 21, 24, 145–146, 166
Networks, 16, 51
Netherlands, 5
Nevada, 37, 39, 80
New Jersey, 37, 201, 203–205, 223
homeownership, 132–134, 137–143, 158–159
laborforce, 100, 102–103

New Jersey *(cont.)*
middle class, 64–69, 72–77, 80–90
professional occupation, 113–119, 121, 123
New York, 21, 25, 31, 36–37, 39, 201, 203–205, 223
homeownership, 132–134, 137–143, 158–159
laborforce, 100, 102–103
middle class, 64–69, 72–77, 80–90
professional occupation, 113–119, 121, 123
New York City, 57, 83, 134, 169, 172
New York Times, 3, 19, 172
Newspapers, 10
Niches, 51, 96–97
Non-Hispanic whites (see whites)
Northern California, 57
Northern Europe, 53
Nurses (see occupations)

O

Occupations
and achievement, 131
administrative, 95
agriculture, 96, 99, 104–105, 214
education, 105–107, 119–121, 124
doctors, 104–107, 111
lawyers, 105–107, 122
managers, 105–106, 119–121
medical, 105–107, 112, 119–121
nursing, 106–107, 121, 124
professions, 78–79, 104–107, 109–119, 121, 124
professional status, 163, 171, 176–177
retail, 97
service, 104, 107
white collar, 104, 177–178

P

Philippines, 36, 49, 118, 133, 151, 182
professional occupation, 108–109, 112
Pluralism, 169
Politics, 164, 186
Political
assimilation, 163
influence, 191
participation, 19, 164, 170–173, 179–185, 191
parties, 172
process, 172–173, 187
representation, 185–188
tension, 166
Polarization, 18, 196
Population
growth, 219–221
Projections, 220
Poverty (poor), 6, 18, 20, 23, 50, 52
and learning, 209
Professions (see occupations)
Propaganda, 10
Protected groups, 186
Public assistance, 20
Public schools, 211, 216

R

Race, 45, 169, 171, 189
Racism, 170
Recessions, 200
Redistricting, 188–191
Refugees, 15–16, 49, 147, 219, 222
Regions, 80, 113
Religion, 35, 165, 168
Remittances, 14, 148
Renters, 156–157
Residential mobility, 156
Retirement, 9
Rhode Island, 32, 44
Russia (Soviet Union), 2, 22, 36, 42, 82, 182
professional occupations, 108, 112, 117

S

San Francisco, 38, 89, 147, 149–154
San Jose, 147
San Joaquin Valley, 213–214
Schools, 25, 44–45, 113, 129–130, 209–211, 216–217
Scientists, 105
Seasonal workers, 22
Second generation, 3, 76–77, 167, 214
voting, 183–185
Self employment, 97
Service (workers), 21, 101, 104
Silicon Valley, 3, 111, 147
Skills, 51–52
Transferability, 4, 57, 211
Soccer, 24–25
Social contacts, 166
Social mobility, 17, 23
Social safety net, 57
Social security, 220
Social status, 95
Socio-economic status/progress, 128
South America, 43, 48

Southern California, 15, 24, 57, 136, 166, 191, 211–212
Southern Europe, 53
Spanish language, 169, 229
Suburb, suburban, 23, 145, 155–158, 160, 212–213
Sweatshops, 17, 21, 25
Sweden, 210–211

T

Texas, 21, 31, 39
  laborforce, 100, 102–103
  homeownership, 132–134, 137–143, 158–159, 187
  middle class, 64–69, 72–77, 80–90
  professional occupation, 113–119, 121, 123
Thailand, 19
Third Generation, 20, 76–77, 167–168, 185, 214
Trade, 56
Transnationalism, 169
Travel, 169

U

Underclass, 20
Undocumented migrants, 15, 53, 124, 168, 172, 175
U.S. Constitution, 165

V

Veterans, 127
Vietnamese, 36, 49, 100, 118, 151, 182
Visas, 55, 219
Voting (voters), 163, 166, 172–173, 179–185
  education, 180
  protected status, 160
  year of entry, 179–180
  second generation, 183–185
Voting Rights Act, 172, 186

W

Wages, 61, 125, 197, 203, 215, 218
  inequality/gap, 51–52, 95
Wall Street Journal, 22
Washington Post, 22
Wealth, 2, 8, 27, 127–128, 130–131, 196, 198, 205, 209
Welfare, 20, 54, 199, 217
Western Europe, 53
White collar (workers), 104–105, 176–177, 179
Whites foreign born, 156, 175, 203–205
  homeownership, 137–143, 183
  middle class, 64–69, 82–84, 86–88
  voting, 183
Whites native born, 69, 82–84, 86–88, 90
  middle class, 64–65, 67–69, 73–75, 84–86, 88, 90
Women, 119–122, 123–124

About the Author

William A. V. Clark, PhD, is Professor in the Department of Geography at the University of California, Los Angeles, where he teaches undergraduate courses on population and the environment and graduate classes on international migration and its outcomes in cities and neighborhoods. He has lectured and taught in Europe, Australia, New Zealand, and Canada and published several previous books, including *Human Migration* and *The California Cauldron: Immigration and the Fortunes of Local Communities*. He was a fellow at the Netherlands Institute for Advanced Studies in the Humanities and Social Sciences in 1993, held a Guggenheim Fellowship in 1994–1995, and in 1997 was elected an Honorary Fellow of the Royal Society of New Zealand. He has an honorary doctorate from the University of Utrecht, The Netherlands.